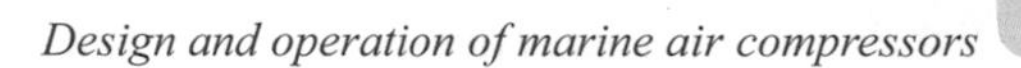

## Acknowledgements

JP Sauer & Sohn Maschinenbau GMBH

Alup Kompressoren

Domnick Hunter Ltd

Zander GMBH

Mobil Oil

Shell Oil

Germanischer Lloyd

Lloyd's Register

And grateful thanks to L Sterling, CEng, author of the original IMarEST book *Selection, Installation and Maintenance of Marine Compressors* (MEP Vol 1 Part 1), from which much of this book has been based.

# About the Author

**Bill Robinson** was educated at Sunderland Junior Technical School, Sunderland Technical College and York Engineering College. After serving an apprenticeship at William Doxford (Engineering) Ltd, he spent several years in the pipe installation and machinery layout drawing offices at both Doxfords and JL Thompsons on the River Wear. He then spent several years in the engineering offices of Swan Hunter Shipbuilders as an engine room equipment estimator before becoming a sales engineer at Compair Reavell covering the sale of marine air compressors.

Once established within Compair Reavell, Bill was promoted through the company before being made Product Manager of the company's reciprocating range of air and gas compressors, including the development and sales of a new range of 90deg 'V' compressors worldwide. These compressors were the forerunner and market leader of compressors with outputs of between 15 and 350 bar delivery pressure. Special attention was given to the naval market where these new air and water cooled compressors were selected by over twenty navies including the Royal Navy. Bill also set up a refurbishment shop within Compair Reavell for the naval range of compressors used by the Royal Navy.

After leaving Compair in 1992, Bill joined the German company JP Sauer & Sohn as Head of UK sales of their air and water cooled range of air and gas compressors for industrial, marine and naval markets. In addition to the UK market, he helped to introduce their warship products to North and South America. These products were modern, state-of-the-art warship machines having their crankshafts sited in the vertical plane and cylinders arranged horizontally giving almost perfect balance and no downward forces resulting in advantages of weight and space saving over their competitors.

When Bill retired at the end of 2002, Sauer compressors were being installed in all new Royal Naval ships and submarines as well as for retrofits and newbuilds on nearly all US navy aircraft carriers. Other US navy ships, eg, LPD 17 were also being fitted with Sauer compressors.

Bill has written and presented several papers to the marine and naval markets including Australia, Brazil, Spain and the UK to name but a few.

# Conversion chart

## *Imperial to Metric/SI Units*

Length
1 Inch = 0.0254m
1 Foot = 0.3048m

Weight
1 lb = 0.4536kg

Airflow
1 $ft^3$/min = 0.0283 $m^3$/min
1 $ft^3$/min = 1.6990 $m^3$/h
1 $ft^3$/min = 28.32 litre/min
1 $ft^3$/min = 0.4719 litre/s

Pressure
1 lb/$in^2$ (psi) = 0.0689 bar

Power
1 hp = 0.7457kW

Heat
1 Btu/h = 0.000293 kW
1 Kcal/s = 4.1868 kW

Torque
1kg(f)m = 9.8066 Nm
1 lb-ft = 1.3558 Nm

## INTRODUCTION

Compression of air has one basic aim and that is to deliver air at a pressure higher than the original. Compressed air is one of the most used forms of energy found in industry today and this has also been observed in the increased number of applications found in today's merchant, naval and offshore vessels. Air systems have several advantages including safety, cleanliness and flexibility.

By far the greatest use of compressed air is for starting diesel engines, normally requiring pressures of between 20 and 40 bar, which is provided by reciprocating air compressors. This type of compressor will have a compression ratio of about 7:1 in each stage depending upon the size, cooling and speed of the machine.

Within this publication, it is intended to cover in a very broad outline, the design, selection and installation of low pressure and medium pressure air and gas compressors, both reciprocating and screw types, used on-board modern merchant ships, cruise liners etc, in order that the machine that has been selected will not run in a high stress condition.

As more and more engine rooms are being designed within the unmanned engine room notation, this area will also be covered in some detail.

Information will also be given on planned maintenance routines as well as condition monitoring which has now been introduced for screw compressors used for working and control air applications. For reciprocating compressors, condition monitoring is still in the early stages of development, but some thoughts on its progress are given.

Another area of importance is the introduction of special compressor mineral and synthetic oils which have significantly contributed to the marked improvement in the reliability of compressors achieved over recent years.

# Chapter 1

# BASIC THEORY OF COMPRESSED AIR

## 1.0 General

All gases deviate from the perfect or ideal gas laws and in some cases the deviation can be rather extreme. It is therefore necessary that these deviations are taken into account in order to prevent the compressor and prime mover size being greatly in error.

## 1.1 Air

A 'perfect' gas obeys Boyle's law (PV = $C$), Charles Law ($V/T = C$) and the Combination Law ($PV/T = C$). Air is a mixture of several gases and the two major constituents that are present are nitrogen ($N_2$ = 78%) and oxygen ($O_2$ = 21%) with 1% of all of the remaining gases by volume. 76% of $N_2$, 23% of $O_2$ and 1% of the other gases by weight. The percentages given are only approximate.

Although not 'perfect' gases, the majority of gases in the air and therefore air obey the above laws very closely. Water vapour will also be present in variable amounts but this will only affect the calculations very slightly and can be ignored for the purposes of this book. In the 'combination' law it is usual to have the constant as a gas constant for the unit, times the mass itself and so enabling the different gas constants to be readily used from the appropriate gas tables. For a given unit mass ie, $PV = RmT$, where $R$ is the constant per unit mass (ie, the familiar $R$ for air is 286 Nm/kg/°K; therefore the 'Combination' Law or characteristic equation for air is $PV = 286$ mT where $P$ = N/m$^2$ absolute, $V$ = m$^3$, $m$ = kg and $T$ = °K.

### *1.1.1 Compression*

It should be noted that both the temperatures and the pressures must be in 'absolute' units.

### *1.1.2 Isothermal Compression*

This law would represent the least energy input and will involve no temperature change, ie, $PV = C$ (Fig 1).

### *1.1.3 Adiabatic Compression (sometimes called Isentropic)*

This is the maximum energy input (Fig 1) which will take place if no heat losses does occur and then this will then become $PV^{\gamma} = C$.

$$\gamma = \frac{C_p}{C_v} \text{ (for air = 1.405)}$$

where $C_p$ and $C_v$ are the specific heats of air at constant pressure and volume respectively.

Actual compression lies between the above two extremes and the law is $PV^n = C$ (n will vary for each type of air compressor but a typical figure is suggested as 1.3). This is known as polytropic compression.

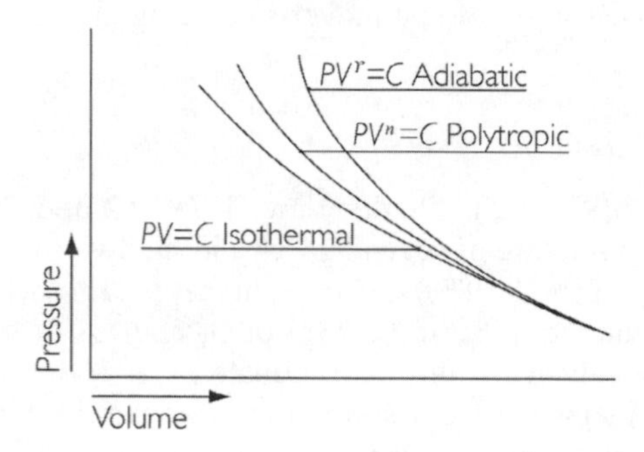


Fig 1: Compression lines

### 1.1.4 *Theoretical Work Done*

**a) *PV* = *C*, Isothermal compression**

Fig 2 shows the compression curve from the volume of $V_1$ to $V_2$. At the point where the volume is $V$, let the pressure be $P$. By allowing a small decrease in the volume $V$ and then the pressure $P$ will be increased by a very small amount.

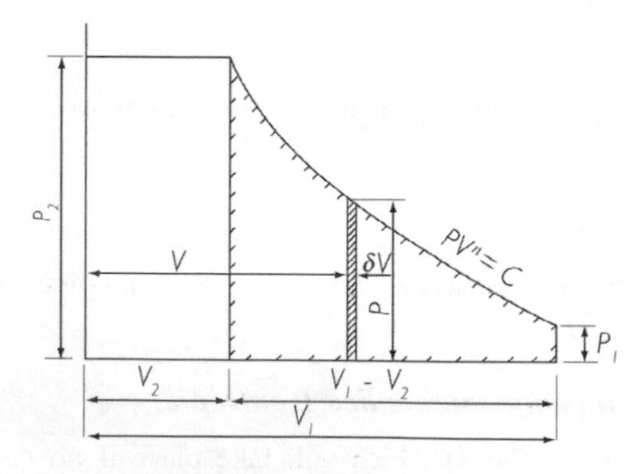


Fig 2: Theoretical work done

Hence, work done = $PdV$
therefore $dW = PdV$

Total work done, $W = \int_{V_2}^{V_1} PdV$ now $C = PV$

$$= C\int_{V_2}^{V_1} \frac{dV}{V}$$

$$= C\left[\log_e V\right]_{V_2}^{V_1}$$

= C (loge $V_1$ – loge $V_2$) now $C = PV$

= $PV$ (loge $V_1$– loge $V_2$)

= $PV$ loge $V_1/V_2$ compression ratio = $V_1/V_2 = r$

Compression ratio $r$ is the ratio between the cylinder volume above the piston at bottom dead centre and the volume above the piston at top dead centre.

$W = PV$ loge $r$ $PV = Rm\ T$

$W = RT$ loge $r$ for unit mass of air

**b) $PV^n = C$, Polytropic compression**

$W = \int_{V_2}^{V_1} PdV$ now $P = C/V^n = CV^{-n}$

$$= C\int_{V_2}^{V_1} V - dV$$

$$= C\left[\frac{V^{-n+1}}{-n+1}\right]_{V_2}^{V_1}$$

$$= C\left(\frac{V_1^{1-n} - V_2^{1-n}}{1-n}\right)$$ now $C = P_1V_1^n = P_2V_2^n$

$$= \frac{P_1 V_1^n V_1^{1-n} - P_2 V_2^n V_2^{1-n}}{1-n}$$

$$W = \frac{P_1 V_1 - P_2 V_2}{1-n}$$

$$= \frac{P_2 V_2 - P_1 V_1}{n-1}$$ now $PV = RmT$

$$= \frac{R(T_2 - T_1)}{n-1}$$ for unit mass of air.

**c) Work done in a cylinder**

Fig 3 shows the theoretical compression diagram.

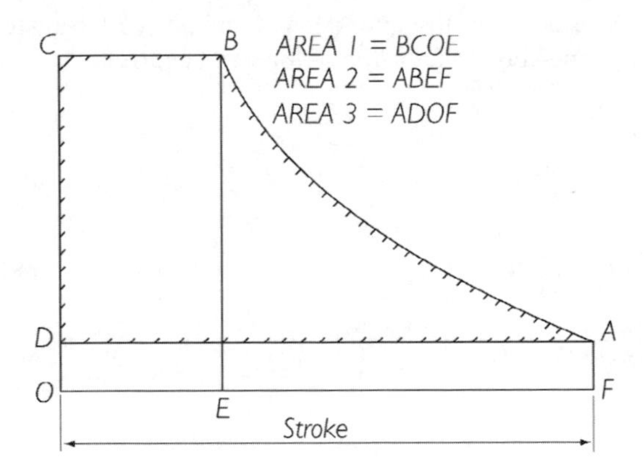


Fig 3: Theoretical work done in cylinder

Work done $W$ = Area 1 + Area 2 - Area 3

$$= P_2 V_2 + \left[\frac{P_2 V_2 - P_1 V_1}{n-1}\right] - P_1 V_1$$

$$= (P_2V_2 - P_1V)_1 \left[1 + \frac{1}{n-1}\right]$$

$$= \frac{n}{n-1}(P_2V_2 - P_1V_1)$$

$$= \frac{n}{n-1}R(T_2 - T_1)$$ for unit mass of air.

## 1.2 Effect of Multi-Stages

Apart from the practical considerations of keeping the maximum temperature of compression compatible with thermal distortion and lubricating oils, etc, power savings can be achieved as shown in Fig 4.

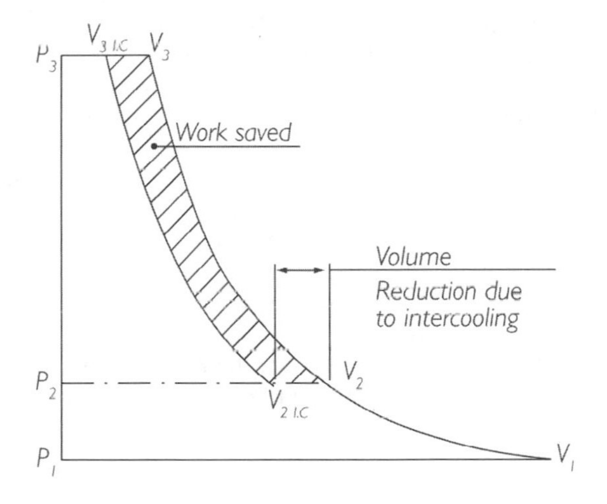


Fig 4: Work saving – multi-staging

## 1.3 Volumetric Clearance

Fig 5 shows the effect of volumetric clearance on the 'perfect' pressure/volume diagram. The volumetric clearance not only lowers the possible output of a given machine but also increases the cylinder charge temperature slightly since the temperature does not drop to the inlet temperature on expansion.

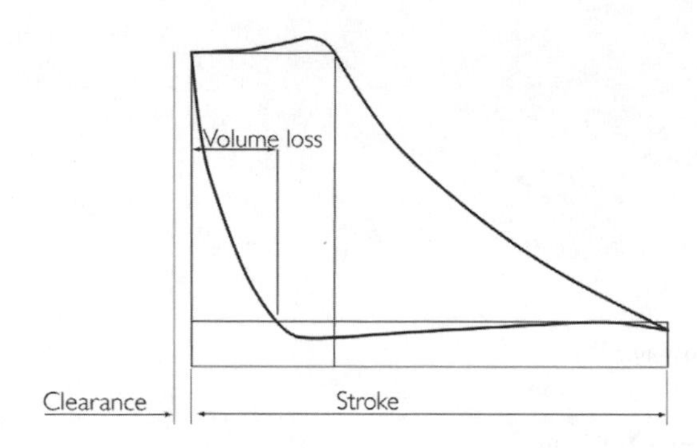


Fig 5: Volumetric loss – clearance

## 1.4 Moisture Effect

The humidity of the air is usually expressed as a percentage and one can reasonably assume an average figure of 70% humidity. The air must be 100% humid before it is fully saturated and free moisture appears. A compression ratio of only 1.43 would be possible before moisture fall out would occur, from 70% humidity air if the compression was isothermal.

With increasing air temperature the ability of air to hold moisture will increase, it actually doubles for every 15°C of temperature increase. Since the temperature increase during compression is considerable, the ability to hold the moisture increases and usually the relative humidity of the air actually increases despite the fact that the air volume is decreasing and humidity increases in direct proportion to the compressor ratio if the temperature remains constant.

On cooling both in the aftercooler and in the pipeline, the ability of the air to retain the moisture volume drops and free moisture will be deposited leaving the compressed air 100% humidity.

## 1.5 Efficiencies

Several types of efficiencies should be considered in compressors, each having their own particular merits.

### *1.5.1 Volumetric Efficiency*

For the considered compressors this is a ratio of the 'free' air compressed and delivered at the compressor discharge divided by the displacement of the compressor. 'Free' air is the air at equivalent condition to the air at atmospheric condition within the engine room.

A high volumetric efficiency (assuming minimum leakages), ensures that the delivery air temperature before coolers is at a minimum for the particular operating conditions. This is because there is less hot compressed air to heat up the new air charge entering and normally the total package (unit) size is also at a minimum.

### *1.5.2 Compression Efficiency*

This is the ratio of the theoretical work of compression divided by the actual work of compression (indicated power).

The theoretical work of compression can be either adiabatic for 'Adiabatic compression efficiency' or isothermal for 'Isothermal compression efficiency'. Since the air will normally be cooled before usage the isothermal compression efficiency is recommended. This efficiency has a direct bearing on the power consumption – see 1.5.4.

### *1.5.3 Mechanical Efficiency*

This is the summation of all of the power losses from the bearings, piston rubbing losses, glands etc, giving the total 'rubbing' losses of all of the machines sliding surfaces. The ratio of the indicated power divided by the actual absorbed power gives this mechanical efficiency.

### *1.5.4 Overall Compressor Efficiency*

This is the product of both the compressor efficiency and the mechanical efficiency.

$$Overall\ compressor\ efficiency = \frac{Isothermal\ power}{Absorbed\ power}$$

### *1.5.5 Overall Efficiency*

This is simply the product of the overall compressor efficiency, the driver efficiency and the transmission efficiency.

# Chapter 2

# COMPRESSOR SELECTION

## 2.0 General

It is very important that the correct compressor type is selected for the correct air application onboard ship. If the incorrect compressor type is selected or sized, this can have a serious effect on the vessel's capability, especially if the units are for starting and stopping the main propulsion engines in manoeuvring conditions and where the safety of life at sea notation and rules play a very important decision making role.

## 2.1 Two Methods of Compression

Air and gas can be compressed by two basic methods:

### *2.1.1 By Compression*

Pressure is imparted by decreasing the air or gas volume using positive displacement machines. Fig 6 shows the four groupings of these main air compressor types.

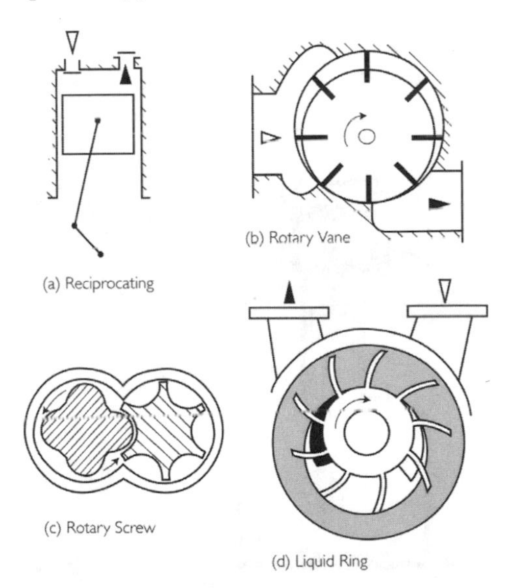


Fig 6: Various Types of Air Compressor

a) Reciprocating compressors are positive displacement machines in which the compressing and displacement element is a piston which has a reciprocating motion within a cylinder. The basic reciprocating compressor element is a single cylinder, compressing on only one side of the piston. A unit compressing on both sides of the piston consisting of two basic single acting elements operating in parallel is known as double acting. On the compression stroke the compressor rises to just above the discharge pressure. A spring loaded non-return discharge valve opens and compressed air is discharged at approximately constant pressure. At the end of the stroke the differential pressure across the valve traps a small amount of high pressure air in the clearance space between the piston and the cylinder head. On the suction stroke the air in the clearance space expands, its pressure drops until such time as a spring loaded valve re-seats and another pressure stroke begins (Fig 7).

b) Screw compressors are rotary, positive-displacement machines with two intermeshing rotors each with a helical form.

c) Rotary vane types are positive-displacement air machines in which both compression and displacement is effected by the positive action of the rotating elements.

d) Liquid ring compressors are rotary displacement machines in which water or the liquid is used as the piston to compress and displace the air handled.

In today's merchant fleets, reciprocating and rotary screw air compressors are the main designs being utilised, with very few of the rotary vane and liquid ring designs now being selected. However, there are several navies still using the vane types of compressors for special air onboard ship applications – these compressors will not be covered in this publication.

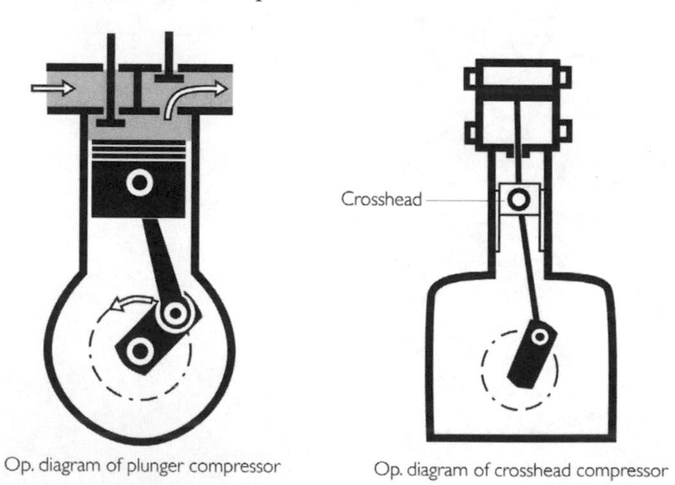


Fig 7: Cycle of Operation

### *2.1.2 By Kinematic Energy*

Dynamic compression (usually rotary vane or lamellar type compressors) imparts kinetic energy to the gases, which in turn is converted into pressure by means of a diffuser. In the marine environment, this group of machines usually contains fans, blowers, superchargers, etc, and therefore will not be covered in this publication.

## 2.2 Choice of Correct Machine

### *2.2.1 Main Start Compressor*

Compressor selection is very important to the operator, since a machine which has been incorrectly selected or has been used for duties other than its design purpose will often cause endless amounts of maintenance problems. Take for example a reciprocating compressor used for main engine starting.

Such an air compressor does not log a great number of running hours per annum and usually will only run for short periods at a time except for periods when the vessel is on passage during a long and complex navigation. The compressor can therefore be selected in its maximum stress condition. This will normally denote selecting the compressor at its maximum running speed and under these operating conditions the compressor will operate both satisfactorily and reliably.

Should the compressor unit be used for additional duties, ie, working air or control air requirements, so that its running hours become longer, a higher stress condition is now created within the machine which will make it necessary for more frequent maintenance periods and possibly will be the cause of more breakdowns.

Even when the duty is for starting air only, the class of vessel will have a bearing on its selection. A starting air compressor used on a short voyage such as a rapid turn round ferry, is in much more constant use, ie, 18 to 20 hours per day and therefore the compressor must be selected in a lower stress condition or be designed to operate under these conditions. At the other end of the scale, the control air screw compressor is designed for a 24 hour day continuous duty and will be selected so that the stress conditions are low for the type of machinery in question.

An alternative is to use a larger capacity starting air compressor, but operating at a lower running speed. A further consideration must be the design age of the compressor itself. Many of the 'older' designs are over 30 years old and have been stretched in some cases beyond their original design limitations. It is therefore the responsibility of the system designer to ensure that the compressor selected operates for the duty that the system has been designed for. Modern compressors over the past 10 years will not have this problem, as they have been designed for higher synchronous speeds ie, 1760 rev/min and have as yet not been stretched beyond their safe operational limitations.

### 2.2.2 Topping-Up Compressor

One of the most abused air compressors installed onboard ship is the so called 'topping-up' compressor, which because of general service air usage taken from the air receivers, plus any system leakage, often becomes a 24 hour per day category compressor. Its design operating parameters are low and its selection is often based on a compressor which does very little running. This may lead to a higher maintenance level than originally planned and expected.

The examples given above are for reciprocating compressors operating at pressures of up to 40 bar. If both separate control air and general service air compressors are to be installed, normally one or more screw compressor will be selected and in the majority of cases the unit will be designed for a 24 hour per day duty cycle and therefore selection under this operating condition will not be a problem.

### 2.2.3 Black Start Compressor

The black start, also known as the emergency compressor, is according to the survey authorities' rules a self-contained unit. In the majority of cases its prime mover will be a diesel engine but under certain conditions the compressor can be fitted with an electric motor as the prime mover.

When sizing the compressor unit, consideration must be given to ensure that under the emergency condition, the unit has to be able to provide sufficient power to start the essential services including the main propulsion within 30 minutes.

As an alternative to a diesel or electric motor driven black start compressor, consideration can be given to a hand operated unit in which the charging lever can be operated at around 50 double strokes per minute. The unit can be complete with a small air receiver with sufficient capacity to give one start to the emergency generator. This emergency system meets the basic requirements of the survey authorities.

A further alternative is if three main start air compressors are installed with power of say between 20 and 25 kW, to utilise one of these main starting units powered from the emergency shipboard with star-delta start.

## 2.3 Class of Compressor

### *2.3.1 Rotary Screw or Reciprocating?*

There is no question in anyone's mind that above the operating pressure of 13 bar, the reciprocating air compressor with its more positive sealing is the only correct machine for selection. The reciprocating compressor limitation is the high temperatures caused by interstage compression, where the compression ratio can be as high as 7 to 1 within each stage, with intercoolers fitted between each of the stages and an aftercooler fitted after the final stage. Therefore a starting air compressor with operating pressures of up to 40 bar gauge can be attained by selecting either a two or three stage machine, depending upon the cooling medium selected.

The rotary screw compressor, which is a single-stage machine, is limited by the amount of gas slip past its seal, to a maximum operating pressure of about 13 bar. Commercial considerations for multi-staging a screw compressor to meet an operating pressure of up to 40 bar required for main starting air, would bring many technical problems especially in the area of shaft sealing. On the commercial side, the relatively small market size together with high development costs have precluded processing this type of design which may have to include higher pressures of up to 400 bar. In years to come the pressures may become even higher with operating pressures of up to 600 bar or even higher being considered, especially in the breathing air market for the professional diver and firefighter.

### *2.3.2 Oil Free and Non Oil Free Compressors*

Normally air for main start air compressors is drawn directly from the engine room via a air suction filter and very rarely is oil free air called for. The air that is drawn into the machine will be of atmospheric humidity and the compression ratio will always be sufficient to produce free moisture, leaving the compressed air 100% humid.

For certain sensitive applications, air may be drawn from a clean area, direct to a flanged air filter fitted onto the compressor inlet.

The moisture produced by compression can cause corrosion, both within the machine itself and also in the air system downstream of the compressor. Therefore the small amount of oil that will be in the air from a non oil-free machine environment will give some protection to the air system. With good machine maintenance the lubricated machine will be the correct choice of compressor except for a system where oil free air is essential, ie, control (instrument) air.

### *2.3.3 Instrument or Control Air*

For this application, a lot of attention has to be paid to the cleanliness and dryness of the instrument (control) air. In the 1970s clean, dry air was in the majority of cases achieved by taking main start air (30 bar), reducing it to the required pressure of between 7 and 13 bar, then passing it through oil removal filters, then an air dryer, to achieve almost 100% clean, dry, oil free air. This method was very expensive and a waste of energy, as the main start compressors (for design restraints see Choice of Correct Machine 2.2) had to be sized to take the total air requirement for both the main start and control air requirements and in some cases, air for working air usage would have to be included within the calculations. In addition, reducing valve stations, to reduce air pressure from 30 bar to say 10 bar would have to be installed within the control air system.

Today it is more accepted for the control air requirements to be provided from a designated compressor, most probably an oil free screw machine operating at a pressure of between 7 and 13 bar, with the final delivered air passing through an adsorption or refrigerant type air dryer unit, to deliver the clean, dry, oil free air.

A cheaper alternative, but almost as effective, would be to install a lubricated screw air compressor, but in addition to the dryer an oil removal filter package would have to be fitted between the compressor and dryer, giving an oil carryover before the dryer of less than 0.1 mg/m$^3$. The oil removal filters would be in addition to the dust removal filters that are provided with the dryers. The compressor providing control air could also be sized to provide air for the general service air requirements so doing away with the need for a separate general service compressor unless a large quantity of air is required.

For a detailed section on screw designed air compressors used for instrument and general service compressors see Chapter 3, Section B.

## 2.4 Sizing the Compressor

The main start air compressors free air delivered (FAD) must be sized in accordance with the rules and regulations of the classification society that the vessel will be built to and the calculation will be as follows:

$$Q = \frac{60 \times Air\,Receiver\,Volume \times (P_2 - P_1)}{1000 \times Time}$$

Where:
Q = compressor size in m$^3$/hour
Air receiver volume measured in litres
$P_1$=lower receiver pressure (as adjusted by shipbuilders) in bars

$P_2$ = higher receiver pressure (as adjusted by shipbuilder) in bars
Time measured in minutes to raise receiver pressure from $P_1$ to $P_2$.
The constants 60 and 1000 are conversions to hours and $m^3$ respectively.

To meet class requirements for safety of life at sea a minimum of two starting air compressors of about equal size would be installed. If the compressors have been designed for other duties in addition to starting air, eg, control air and/or general service air, then this additional air requirement must also be added to the air receiver capacities. The main air start compressors should be sized accordingly.

## 2.5 Sizing the Motor

The power requirement of any compressor is the prime basis for sizing the prime mover and for the selection and design of the compressor components. The actual power requirement is related to the theoretical cycle through the compressor efficiency which will have been determined by test on prior machines. Compressor efficiency is the ratio of the theoretical to the actual air kW power and does not include mechanical frictional losses.

For air compressors with an alternating current drive, the power input may be calculated from the voltage input respectively from the electrical power input of the motor.

The power input may be calculated when the voltage V and current I or the output of the electrical motor is measured.

The power input (motor output) of the compressor is:

$$P = \frac{V \times I \times \sqrt{3} \times \cos\phi}{1000} \times \eta$$

Where:
P = motor output power in kW
V = voltage measured at the motor
I = current in amps measured at the motor
$\cos\phi$ = motor power factor
$\eta$ = motor efficiency.

Power factors and motor efficiencies may be found in the electric motor manufacturer's test data.

Converting the measured values into contract values:

$$p_v = P \times \frac{N_v}{N_g} \times \frac{p_{1a}}{p_a}$$

Where:
$P_v$ = contract power in kW
P = input power in kW
$N_v$ = speed in rev/min as per contract
$N_g$ = measured speed
$P_{1a}$ = contract inlet pressure in bars
$P_a$ = measured inlet pressure in bars.

The power requirement of a compressor is the prime basis for sizing prime movers and for the selection and design of the compressor components. The actual power requirement is related to the theoretical cycle through a compressor efficiency which has been determined by test on previous machines. Once the theoretical power calculation has been determined several compressors of similar output are run to establish the actual power absorbed so that this can be included in the manufacturer's catalogue.

## 2.6 Applications

The number of compressed air applications onboard modern merchant and offshore vessels is increasing and may now include:

Main start air
Topping up air
Emergency (black start) air
Instrument (control) air
Working (service) air
Breathing air
Seismic air.

Breathing air compressors have to be designed to operate at pressures of up to 350 bar and are used for filling small portable air flasks carried on the back of firefighters. Although these compressors are of the reciprocating type this application has not been included in this publication.

Other specialist applications such as seismic air and cement blowing used in the offshore markets have not been included in this publication.

## 2.7 Air Cooled versus Water Cooled Compressors

Technically there is no difference in the final performances of these two designs of reciprocating machines apart from the following observations:

## *2.7.1 Air Cooled Compressors*

Air cooled machines do not have the corrosion problems, nor intercooler fouling problems to anything like the same degree as that of the water cooled machines (except closed circuit, fresh water cooled machines). However, the two stage, air cooled machines will tend to run hotter and radiate more heat into the areas where they are installed. Due to this extra heat, both the two or three stage air cooled compressor units will require air trunking to be installed nearby to direct the cooling air across the machine.

Because of these higher temperatures (up to 55°C), for larger capacity units, three stage air cooled compressors have been introduced. Spreading the stage pressure over three stages instead of two provides the advantages of much lower compression temperatures. This in turn reduces the amount of carbon formation within the machine, thereby providing much improved reliability over a water cooled machine. In the majority of installations where engine room temperatures are relatively low no additional cooling air trunking is necessary. If high temperatures are anticipated cooling air trunking should be installed to direct cool air across the unit and Fig 8 shows the correct air flow arrangement across a unit.

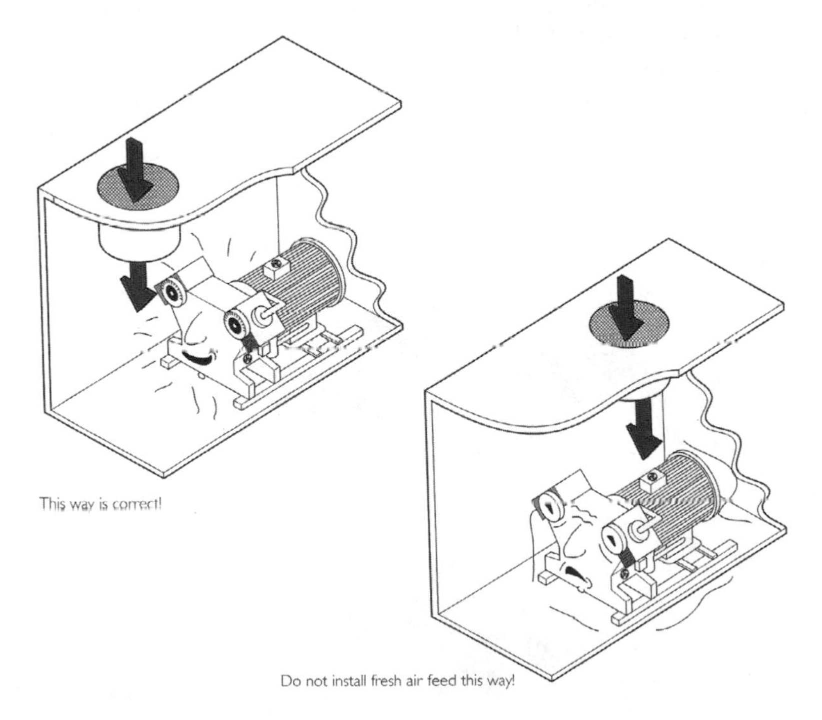


Fig 8: Cooling Air Flow

Although past experience suggests that the water cooled compressors are still preferred by the majority of superintendent engineers, the new generation of two and three stage air cooled machines, with lower installation costs, easier maintenance and longer component life, are rapidly proving that these compressors are the machine of the future on all types of marine vessels.

Further advantages are less weight, reduced space and lower number of pipework interfaces.

### *2.7.2 Water Cooled Compressors*

For many years the use of water cooled compressors have become a selection tradition for main start air application. Their reliability is known and in many cases the same compressor type is used repeatedly in a series of vessels in order to reduce the spare parts holding for the shipowner. Seawater cooling will also help to reduce interstage temperatures, but if fresh water cooling is utilised, the interstage air temperatures will be several degrees higher.

Water cooled compressors are more prone to crankcase condensation especially in very cold cooling water operating conditions, however the formation of carbon can be much less on a modern 90deg 'V' unit, compared to a vertical two stage, water cooled machine. In general the noise level will also be slightly lower than on an air cooled unit due to the deadening effects of the water jackets on the machine.

One disadvantage of seawater cooling is that the high salt content may attack the compressor tubes, cylinder walls and cooling water pipework. It is necessary to fit zinc or similar anodes within the cooling jackets for corrosion protection.

Fig 9: Water Cooled Compressor

Since the introduction of air compressors with the cylinders arranged in a 90deg 'V' or 'W' configuration and use of special mineral or synthetic oils, reliability of the water cooled machines have improved considerably making this type of machine much more acceptable as an operational unit. Fig 9 shows a typical 90deg 'V' configuration main start air compressor.

### *2.7.3 Summary*

Although past experience suggests that the water cooled machines are still the preferred option by shipowners, the introduction of new generation two and three stage air cooled machines, with lower installation costs, easier maintenance and longer component life are becoming increasingly popular.

# Chapter 3

# COMPRESSOR DESIGN

## 3.0 General

This section has been divided into two sections, 'A' for reciprocating compressors and 'B' for screw compressors.

## 3.1 'A' Reciprocating Compressors

Reciprocating air compressors installed onboard have three basic designs: vertical, 90deg 'V' or 'W', and horizontal. The 'V' designs offer several major advantages over the other two namely:

- Generally much less vibration due to the 90deg 'V' or 'W' design of the cylinders.
- A shorter crankshaft also results less vibration and is a more sturdy design.
- More compact design ie, less length and height.
- Maintenance is simpler because of easier access.
- Shorter stroke allows for much higher rotational speeds.

Because of these advantages, the majority of compressor designers are today favouring the 90deg 'V' or 'W' configuration. These compressors are of the two stage or multi-stage types which have the following advantages:

a) saving power

b) reduced air discharge temperature

c) to limit the interstage pressure differential.

1$^{st}$ stage pressure on two stage machines is about 6 bar gauge and on three stage machines the 1st stage pressure is about 2.2 bar gauge and 10 bar gauge entering the 3$^{rd}$ stage. Most two stage machines have a final pressure of 30 bar gauge, whereas three stage units will have no trouble obtaining pressures up to 40 bar gauge.

Because of intercooling between compression stages there will be a reduction in the maximum air discharge temperature. Limitation of the maximum discharge temperature is particularly important for safety when handling air or in high pressure compressors where distortion of cylinder parts may be a problem.

## 3.2 Valves

The valves of a reciprocating air compressor are the hardest working parts of the machine and may have to operate in many different ambient conditions from below zero to plus 60°C.

The older type of vertical, water cooled, air compressors were originally designed with separate suction and delivery valves, and today there are thousands of units of this type in operation. Since the mid 1980s with the introduction of the 90deg 'V' and 'W' configuration machines, this type of valve arrangement has been superseded by a combined suction and delivery valve, with the suction part of the valve in the centre and the discharge on the outside – nearest to the cooling medium. Fig 10 shows a typical arrangement of this type of valve.

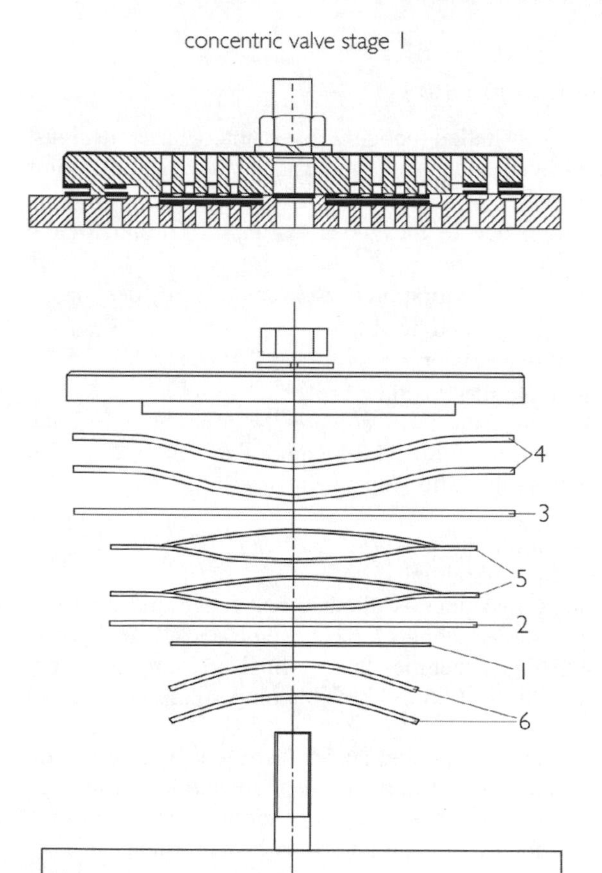


Fig 10: Combined Suction and Delivery Valve

It is important that all types of valve have a good sealing arrangement, to prevent the air pressure from escaping back to the suction side of the valve during the compression stroke. The valve pressure losses, maintenance requirements and clearance volumes, are the main disadvantages in reciprocating compressors not found in other compressor designs.

The operation of the compressor can vary considerably from one type of vessel to another for example some vessels units are operating for nearly 20 hours per day, whilst an oil tanker may require very little compressor usage during the voyage, unless the main start compressors have also been sized to provide the air for the control air or general service air requirements.

Valve failures occur due to wear and fatigue, the presence of foreign particles and overheating within the compressor. Solid particles can be easily detected, but if the contaminants are in a liquid form, ie, water, then this may cause other problems, eg, excessive piston ring wear and damage to compressor lubricating oil. The presence of carbon forming on the valves is a sign that the compressor is running too hot and it is likely to be found on the discharge side of the valve and in the discharge piping. If the carbon becomes brittle it could break off and find its way into the cylinders which could cause serious damage to the cylinder walls and piston rings.

On older designed two stage compressors with separate suction and delivery valves it is normally recommended that valves should be examined between 500 and 800 running hours. On the modern, 90deg 'V', two stage machines which are fitted with combined suction and delivery valves (known as concentric valves), the maintenance period for either a water-cooled or two stage air-cooled machine can be increased to between 1000 and 2000 running hours. The three stage air-cooled machines can run for even longer periods with servicing of the valves required every 2000 hours.

The cleaning or replacing of valves will depend upon their condition. Broken or worn parts must be replaced before valve failure occurs and all components must be carefully cleaned either with a recommended cleaning fluid or with the use of a soft brush. Wire brushes or sharp edged tools must not be used to clean valve seats, springs and plates. Valve seats should be inspected for any ridges, pitting or burning and the extent of the damage, will determine if the valve can be reconditioned or replaced as a complete unit.

After the valves have been reassembled, they should be checked to ensure that they are free to move and will fully open. Valves are usually seated in the cylinder either by metal to metal contact or by using a gasket. If the former is used then the valve should be 'lapped' in place to ensure that there are no marks on the surface that would prevent a leakproof seal. If gaskets or 'O' rings are used, they should be replaced each time the valve is inspected or replaced. 'O' rings have a tendency to expand especially if synthetic oil is used.

Valves are complex parts and should only be opened if there is a problem or for scheduled maintenance.

When replacing valves, the second stage valve is smaller than the first and if the compressor is a three stage machine, the third stage valve will be smaller than the second.

## 3.3 Valve Problems

For more information on valve problems see the Tables at the end of Chapter 9.

## 3.4 Types of Valves

### *3.4.1 Reed Valve*

The reed valve (Fig 11) is also known as a lamellar valve. It is a type in which a reed or flap is positioned over an opening in the valve plate. This type of valve is used extensively in small, air cooled, medium and high pressure air compressors.

Because of its simplicity, little maintenance is required on this type of valve and replacement is preferable to overhaul.

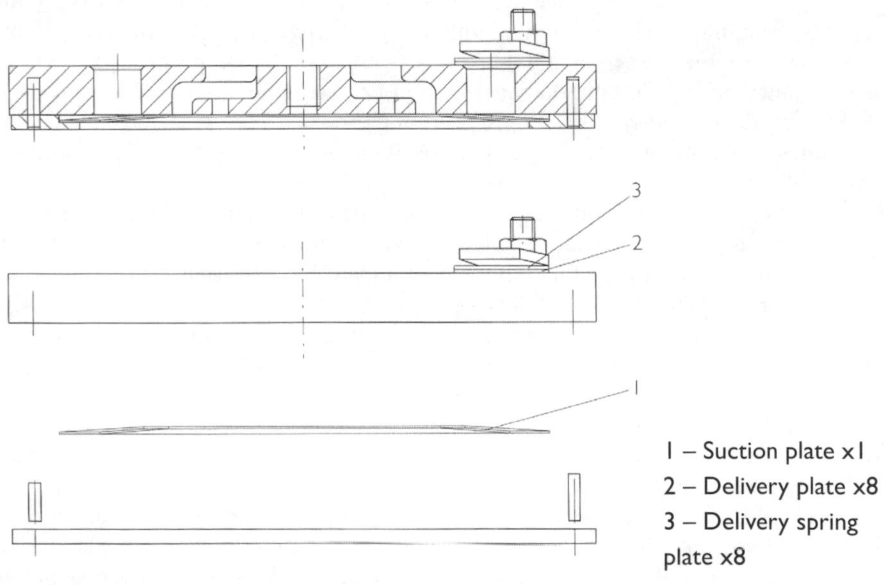


Fig 11: Reed Valve

### 3.4.2 Plate Valve

This is a very common type of valve. It has a number of annular openings through the seat of the valve and the valve plate is located next to this and is designed so that there are blank areas, covering all of the openings in the air passages. Between the blank areas in the plates, there are openings which allow the air to pass when the valve plate lifts and the suction air passes downwards through the centre of the valve. After compression in the cylinder, the delivered air is passed upwards through the outside openings of the valve which are nearest to the cooling medium.

## 3.5 Connecting Rod Bearings

The connecting rod sleeve bearings carry the overall forces applied to the connecting rod bearings. In a vertical, single stage or two stage unit, this would be the force and direction of the connecting rod loads which would be split between the crankshaft main bearings.

In multi-cylinder compressors the cylinders are arranged at an angle, ie, 'V', 'W' or horizontally. The various forces and angles have to be designed to determine the final loads and their directions.

The majority of main bearings used in compressors are designed in two halves and have whitemetal inserts bonded to a steel backing but in newer compressor designs roller bearings are being introduced which are cheaper, easier to replace and have a longer bearing service life.

## 3.6 Connecting Rods

The pistons are connected to the crankshaft by a connecting rod with the smaller bearing at the piston end. The latter bearings are normally of the roller type, while at the crankshaft main bearings are fitted, [see 3.5].

Older designs of compressor (pre-1980) usually have a long connecting rod and the compressor speed will vary between 760 and 1200 rev/min which result in piston speeds in excess of 5 m/s. The modern 90deg 'V' units are designed to run at speeds of up to 1800 rev/min and due to the shorter connecting rod stroke piston speeds of 3 to 4 m/s can be achieved. A shorter stroke connecting rod design helps to reduce valve and piston ring wear.

The material used to manufacture the connecting rods can either grey cast iron or spheroidal graphite cast iron although similar materials may be selected according to the manufacturer's standard.

The piston is connected to the connecting rod small end bearing by a piston pin (gudgeon pin) and it is essential that both small and big end bearings have an adequate supply of lubricating oil. A typical connecting rod fitted with main and small end roller bearings is shown in Fig 12.

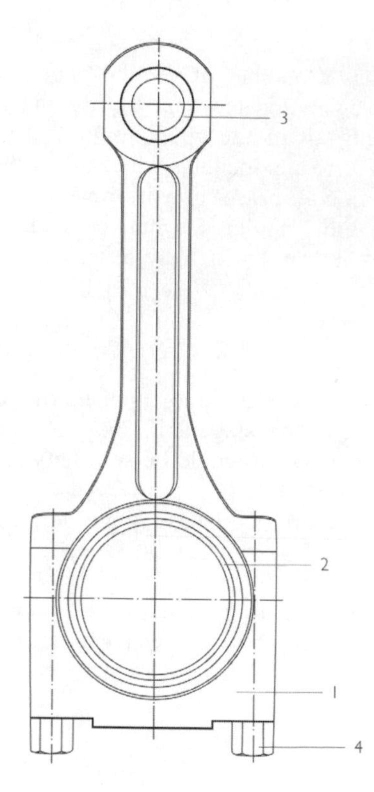


1 – Connecting rod x1
2 – Connecting rod bearing x1
3 – Needle roller bearing x1
4 – Connecting rod screw x2

Fig 12: Typical Connecting Rod

## 3.7 Crankcase

The crankcase will be designed to meet the manufacturer's requirements to house the crankshaft, cylinders etc. Materials are similar to that used for the cylinders, ie, spheroidal or grey cast iron. On the smaller designed air cooled units aluminium crankcases are more likely.

Crankcase doors will be fitted onto the crankcase, normally one on each side, so that visual inspections may be made and to facilitate necessary maintenance.

The bottom of the crankcase should be checked twice annually especially around the drain plug in the sump to remove collected debris. The presence of water and sludge deposits should also be checked in accordance with the manufacturer's instructions.

## 3.8 Crankshaft

The crankshaft design must be of sufficient strength to meet the selected classification society's rules. It will be designed to ensure that the balance weight throws are as large a radius as possible with the machine in its normal bolted down position and then with its feet unbolted. If a bedplate is required any discrepancy between the two sets of readings (crankshaft deflections), will show that there is a misalignment caused by incorrect chocking on the seat mountings.

Present designs make considerable use of computer analysis to determine wide ranges of loadings and dimensions before manufacture. It is important that the crankshaft has an end float to allow for expansion and these figures should always be listed as part of the compressor operating and maintenance manual. If expansion figures are not available, a minimum of 0.025 mm per 25 mm of allocation length should be allowed as a guide.

Main bearing failure rarely occurs but if it does then this may be due to crankshaft misalignment and major servicing would be necessary.

It is possible to design the crankshaft as a cantilever with the bearings at one end only and with the other end of the crankshaft floating. This configuration will reduce the compressor costs but may take away some of the rigidity of having bearings sited at each end of the crankshaft. In addition during major servicing of the unit, the service engineer may experience some difficulty in working on the compressor and may require the assistance of a second pair of hands.

Crankshaft materials will depend upon the manufacturer's standards, grey cast iron, spheroidal graphite cast iron or even nodular cast iron. A typical crankshaft arrangement is shown in Fig 13.

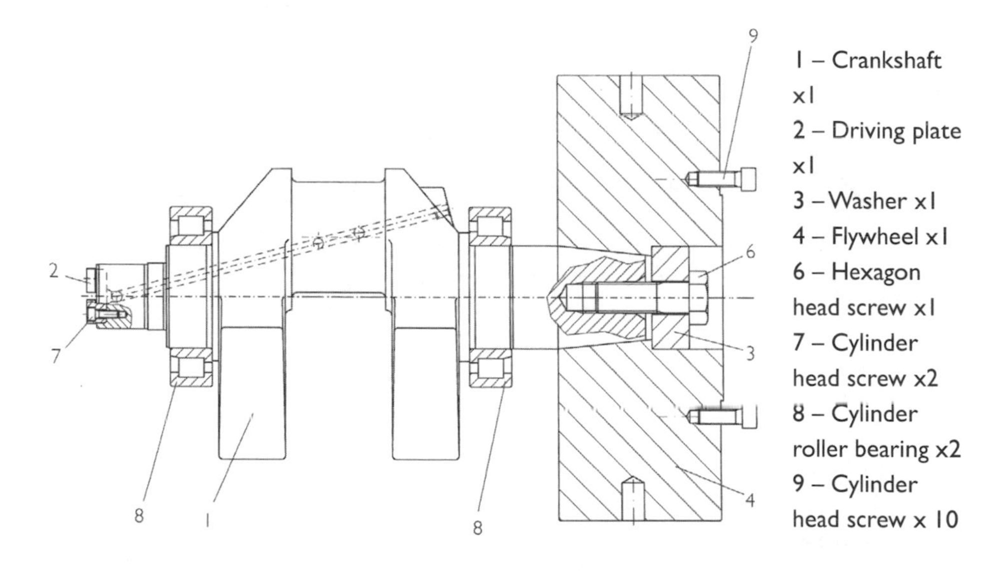


Fig 13: Crankshaft

## 3.9 Pistons

The main function of the pistons is to induce air into the cylinders by transferring energy from the prime mover via the crankshaft to the air which is compressed and then discharged from the cylinder. To ensure that this compression is carried out as efficiently as possible piston rings are fitted to the pistons which make a seal between the piston and the cylinder wall preventing air from escaping into the crankcase. Piston rings also reduce the amounts of oil that pass into the compressor stages and mix with the air in the compressor. For a typical first stage piston arrangement see Fig 14.

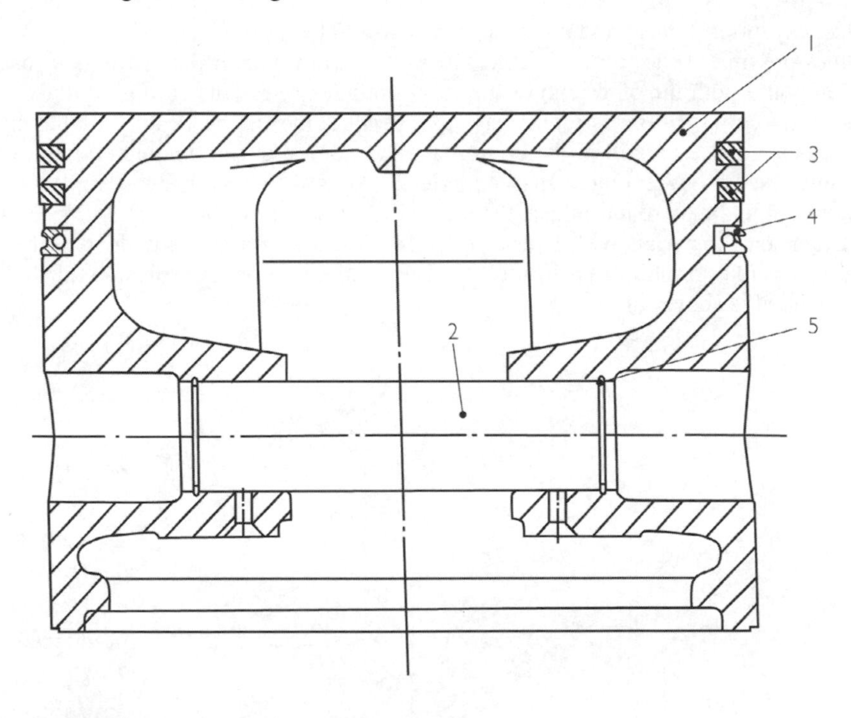


1 – Piston x1
2 – Piston pin x1
3 – R-ring x2
4 – G-ring x1
5 – Circlip x2

Fig 14: First Stage Piston

The pressure applied to the pistons at any one time depends upon the piston area and the pressure difference across it. The acceleration forces are dependent upon the compressor speed and the mass of the components and it is recommended that the maximum piston speed is 5 to 6 m/s.

In the design of the pistons the first stage piston will almost certainly be manufactured in aluminium alloy to give the correct balance over the machine. Cast iron is the material used for second and third stage pistons (if fitted).

### *3.9.1 Piston Rings*

Piston rings are an important part of a compressor design as they provide a seal to prevent the air from passing into the crankcase as well as creating the compression to raise the pressure between each stage.

Piston rings for an oil lubricated compressor are manufactured in cast iron and the ring pack configuration will depend upon the manufacturer's design standard.

As an innovation some manufacturers have introduced plastic piston rings in the final stage although it may be too early to advise of any reliability conclusions covering this new material. One possible problem area that may be highlighted is that if high temperatures are experienced within the final stage distortion may take place on the plastic ring material and cause consequential damage within the cylinder.

The piston rings should be checked and measured in accordance with the manufacturer's recommendations which will be detailed in the instruction manual. There must always be adequate clearance between the piston and the liner to allow for expansion. Always replace all piston rings when one of them exceeds the limit. Install rings on the respective piston and make sure they are in the correct position. Piston rings have an asymmetric cross section and are marked on one of the surfaces with 'Top'.

If an oil free reciprocating compressor is used for control air applications the piston rings will probably be PTFE coated and carbon filled. It will be necessary to install an appropriate filter after the compressor final discharge to trap any dust particles before they are carried downstream into the air system.

### *3.9.2 Liners*

Wet or dry liners may be used to suit the requirements of a compressor design. It is good engineering practice to fit a 'pressed in' liner in each cylinder of a water cooled machine and this can either be of the dry or wet type. The preferred choice is the dry liner which does not require an 'O' ring seal between the liner and the water jacket. This reduces the possibility of corrosion forming on the liner and cylinder if the seal fails or becomes twisted during assembly.

For air cooled compressors, liners are not normally required but if they are fitted they would be of the dry type.

Liner material would be similar to that used for the cylinders and they should be of sufficient thickness to withstand the air pressure.

## 3.10 Cylinder Cooling

Heat in a cylinder results from the work of compression plus the friction of the pistons, piston rings on the cylinder walls, and rod packing on the stage connecting rods. Heat can be considerable particularly when moderate to high pressure ratios are included.

Most compressors use some method of dissipating a portion of this heat so reducing both the cylinder wall temperature and the air final temperature. There are several advantages in cylinder cooling at least some of which apply to all but a few exceptional cases.

a) Lowering cylinder wall and cylinder head temperatures causes a reduction of losses in compressor capacity and kW power per unit volume due to suction air pre-heating during the inlet stroke. There will be initially a greater mass of air in the cylinder ready to be compressed.

b) Reducing cylinder wall and cylinder head temperature will remove more heat from the air during compression lowering its final temperature and reducing the power required.

c) A reduction in air temperature and in that of the metal adjacent to the valves provides a better operating environment for these parts giving longer service life and reduction of carbon deposit formation.

d) Reduced cylinder wall temperature parameters result in better lubrication and consequently longer service life and lower maintenance.

e) Water cooling in particular maintains a more even temperature around the cylinder base and reduces the possibility of distortion.

## 3.11 Water Cooled Compressors

Water cooled, (fresh or seawater) starting air compressors (Fig 15a), have cylinders arranged in a vertical configuration with one first stage and one second stage arranged in line. To increase machine output, it is possible to design a compressor as a block with two first and second stages of compression.

The latest designs of water cooled compressor cylinders are arranged in a 90deg 'V' or 'W' configuration. Both of these unit types may be banked and have up to two first and second stages dependant on the desired output.

Determination of the amount of heat removed by cylinder jackets in specific cases is far from an exact science because of the number of variables. Such factors as: cylinder size; piston speed; air characteristics; cylinder wall thickness inclusive of dry liners; effective area of water jacketing and average temperature difference between air and water will influence the quantity of heat rejected to the water.

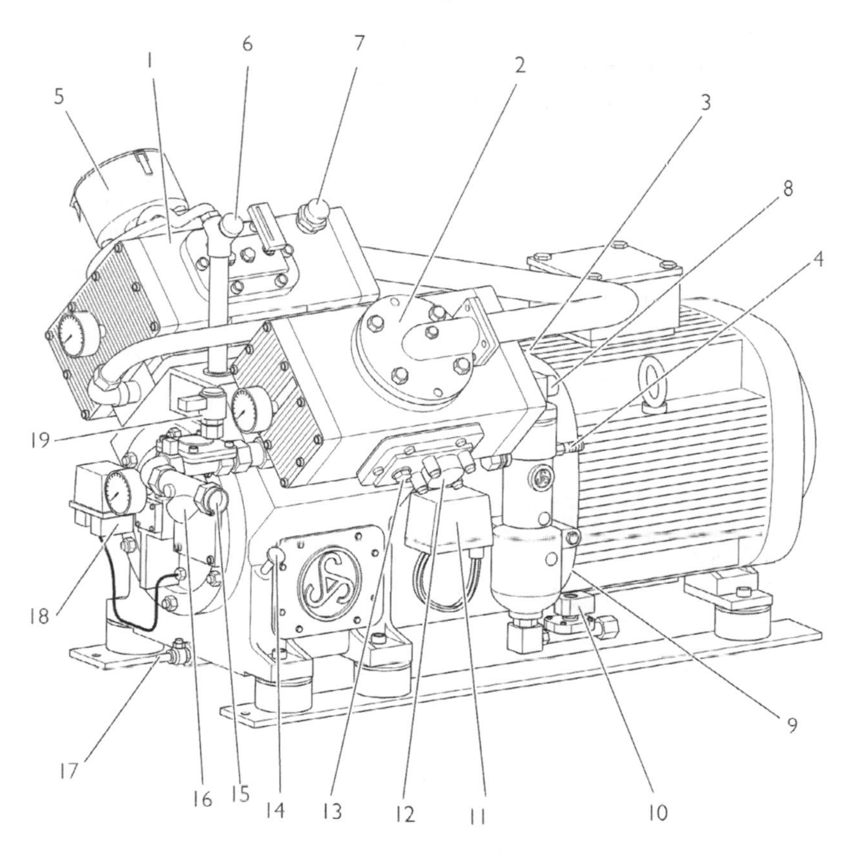


1 – Cylinder 1st stage
2 – Cylinder 2nd stage
3 – Safety valve 1st stage
4 – Safety valve 2nd stage
5 – Air filter
6 – Oil filter
7 – Cooling water outlet
8 – Compressed air outlet
9 – Drain valve 1st stage (not visible)
10 – Drain valve 2nd stage
11 – Compressor air temperature monitor
12 – Bursting disc
13 – Zinc protection 2nd stage
14 – Oil dipstick
15 – Cooling water inlet
16 – Particle trap
17 – Oil drain valve
18 – Oil pressure monitor
19 – Cooling water stop valve

Fig 15a: Typical 90 deg 'V' Water Cooled Compressor

## 3.12 Air Cooled Compressors

The majority of modern, small capacity (up to approx. 80 $m^3/h$), air cooled units have two stages arranged in a 90deg 'V' configuration. Larger capacity compressor units are of 'W' configuration and to reduce cylinder temperatures it is recommended to have three stages of compression. To assist in the cooling of these units the outside of the cylinders and cylinder heads are finned to provide an optimum cooling surface area.

Materials of both types of cylinder will either be manufactured from grey or spheroidal graphite cast iron, dependant on the manufacturer's standard. The smaller air cooled units may be manufactured either from aluminium or cast iron. Fig 15b shows a typical two stage air cooled compressor.

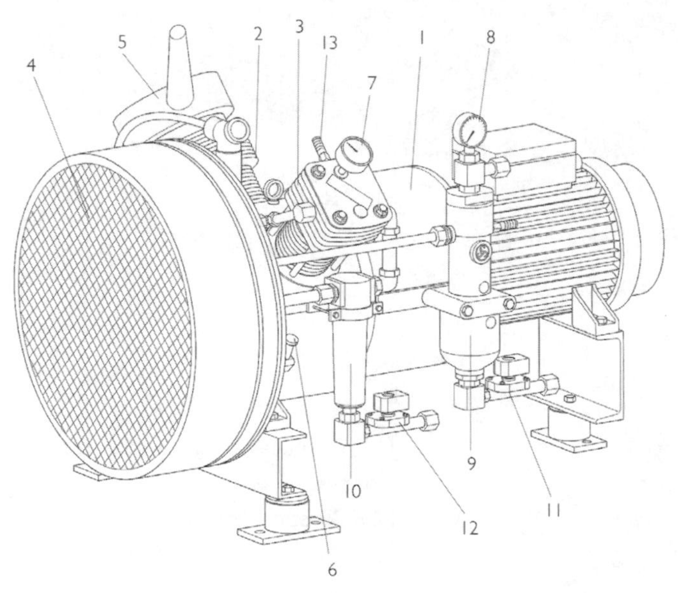


1 – Crankcase
2 – Cylinder $1^{st}$ stage
3 – Cylinder $2^{nd}$ stage
4 – Cowl plus $1^{st}$ and $2^{nd}$ stage cooler coils (hidden)
5 – Air filter
6 – Oil dipstick
7 – $2^{nd}$ stage pressure gauge
8 – $1^{st}$ stage pressure gauge
9 – $1^{st}$ stage separator
10 – $2^{nd}$ stage separator
11 – $1^{st}$ stage drainage
12 – $2^{nd}$ stage drainage
13 – $2^{nd}$ stage relief valve

Fig 15b: Two Stage Air Cooled Compressor

## 3.13 Interstage Coolers

To reduce temperatures created within the compressor it is necessary to install a cooler after each stage of compression and an aftercooler after the final stage. These coolers are installed so that the final delivered air temperature should not be greater than 20°C above the air inlet temperature.

Cooler tubes can either be straight and fixed at each end of the cooler body or they can be of the 'U' type fixed at one end only. The latter is a much cheaper arrangement because of the smaller diameter tubes and the necessity of having only one end cover. The disadvantage of this type of arrangement is that because of the bend the tubes have a tendency to trap carbon that may be present, increasing the air temperature within the compressor, leading to wear and damage to components and at worst leading to explosion and fire. In both cooler designs withdrawable cooler tubestacks rather than fixed cooler tubes are preferable to allow easier maintenance.

Water cooled intercoolers cause few problems providing the water side of the cooler tubes are kept clean. On the air side of the machine if the machine is correctly cooled and there has been no excessive temperature increase caused by valve failure the coolers should remain relatively clean with only a smear of oil present. If the air temperatures become excessive due to valve problems the cooler inlet will become heavily carbonised. The delivery side of the valve will remain relatively free from carbon but show signs of the effects of the heat generated. Carbon formation may also affect operation of the 1st stage relief valve which must be treated seriously and the cooling system of the machine investigated.

On air cooled machines the intercoolers are designed with fins to maximise the cooling surface area and to guide the cooling air that is drawn across the machine by the fan driven directly from the crankshaft.

The function of the intercooler is to reduce the compressed air temperature as low as possible before the air passes onto the next stage of compression. This will bring the air temperature down to below its dew point. It is very important that any condensate present after each stage of compression is collected and drained away from the machine preferably by including an oil and moisture separator after each stage which will also be fitted with an automatic drain arrangement.

Machines fitted only with manual drains have to be operated at intervals of between 15 and 20 minutes dependant on engine room ambient conditions. This system is not the best solution for removing condensate since in the majority of cases after a period of time this operation may be overlooked by the engineer, resulting in condensate finding its way into the sump and into the air system downstream of the compressor.

Cooler tubes are normally manufactured from either copper or cupro-nickel. Fig 16 shows an air cooled compressor in section.

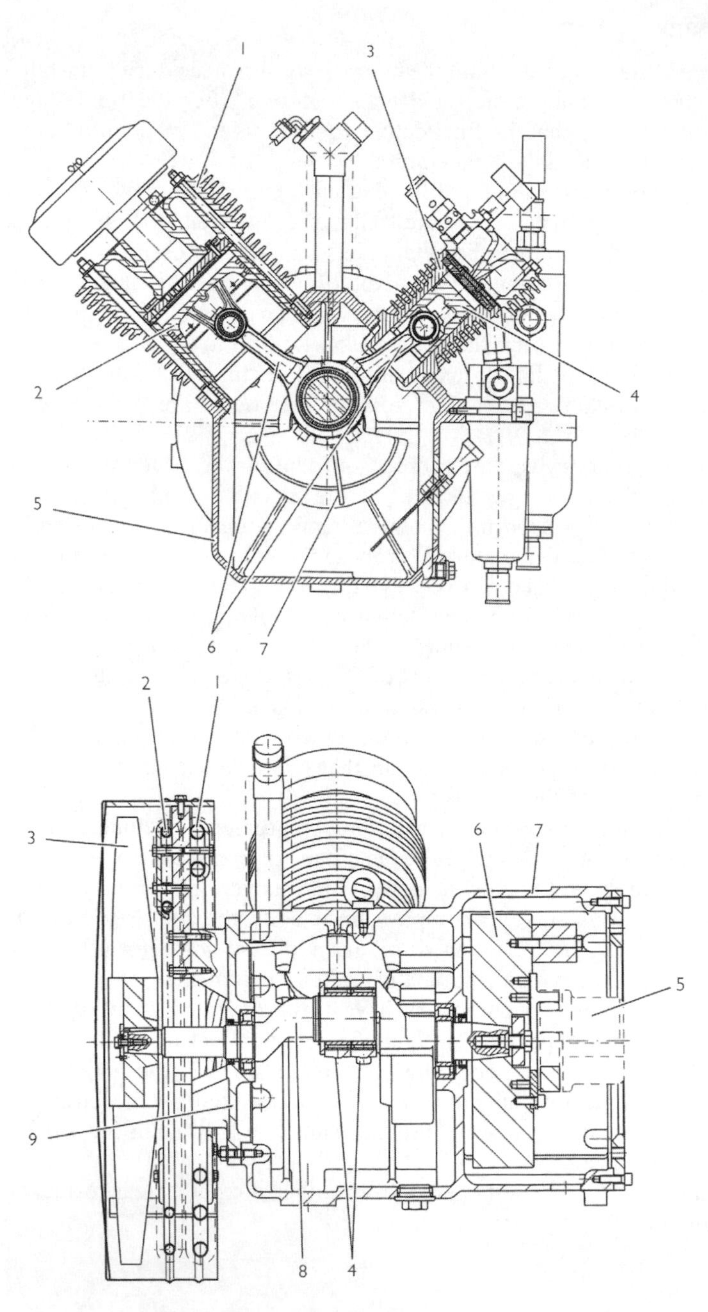


Fig 16: Sections of an Air Cooled Machine

(Top)

1 – 1st stage: cylinder with head and valve
2 – Piston 1st stage
3 – 2nd stage: cylinder with head and valve
4 – Piston 2nd stage
5 – Crankcase
6 – Connecting rod
7 – Dipper

(Bottom):

1 – Cooler 1st stage
2 – Cooler 2nd stage
3 – Fan wheel
4 – Connecting rod
5 – Coupling (flexible for electric motor drives; centrifugal clutch for diesel engine drive)
6 – Flywheel
7 – Transmission bell housing
8 – Crankshaft
9 – Bearing bracket

## 3.14 Aftercooler

To reduce the final delivered temperature of the air to an acceptable level it will be necessary to install a cooler (known as an aftercooler), after the last stage of compression. Both air and water cooled aftercoolers are very efficient and are capable of reducing the final air temperature from around 220°C, to approximately 30°C depending on the ambient conditions.

The water cooled aftercooler body is made in a similar type of material to that of the compressor itself and should be fitted with both a corrosion rod and a bursting disc. As the inlet temperature to the aftercooler can be around 220°C the cleanliness of this cooler is more critical from a safety view point. The cooler has no influence on the compressor performance since its action takes place after the work has been done on the air. The function of the aftercooler is the delivery of the final delivered air at a reasonable temperature which assists in the removal of the condensate. The cooler tubes will be manufactured in a similar material to the intercooler tubes.

On air cooled compressors the coiled fixed tube intercoolers and aftercooler are fitted around the compressor fan. Cool air is drawn across both the cooler coils and the cylinders, to reduce the interstage and final stage air temperatures. The delivered air temperature of an air cooled machine will be approx 5 or 6°C higher than a water cooled unit. If a three stage compressor is utilised the delivered air temperature will be lower compared to a similar sized two stage unit due to the cooling affect being spread over more stages. Fig 17 shows the effect of temperature of a three stage air cooled machines in comparison to a two stage water cooled compressor when the delivered air temperature would be lower, ie, 2 to 3°C above a water cooled machine.

## 3.15 Fusible Plugs/Temperature Switches

On older designs of marine compressors to meet classification society regulations it is mandatory to fit a fusible plug or temperature cut out switch immediately after the outlet of the compressor. This safety feature will protect the system and not the compressor.

On modern designs a temperature switch should be fitted between the final two stages of the compressor which will automatically shut the compressor down if the delivered air temperature rises to approximately 15% above the normal delivered air temperature recommended by the manufacturer. This will allow the air to escape if an excessive air temperature occurs within the machine itself eg, if a cooler tube started to leak. The fitting of such a switch will protect the compressor.

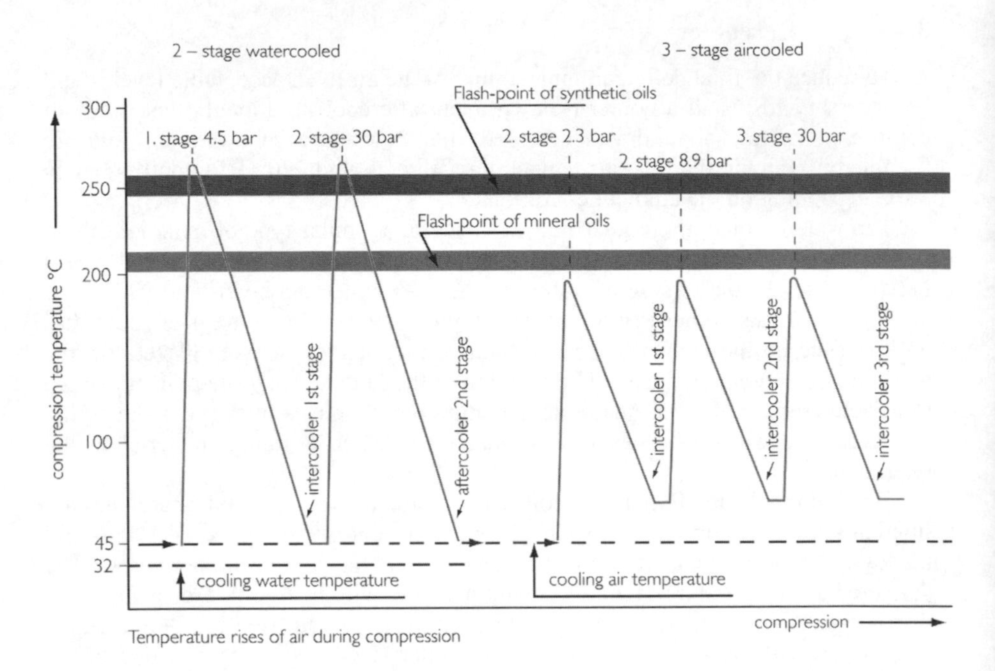


Fig 17: Temperature comparison between 3-stage air cooled machine and 2-stage water cooled compressor

## 3.16 Thermometers

The fitting of thermometers or temperature transducers after each stage of compression and at the water inlet and outlet connections are not mandatory but it is good engineering practice so that the behaviour of the air and water conditions within the compressors can be monitored.

## 3.17 Safety Valves

It is mandatory for a relief valve to be installed after each compression stage and located where the air may become trapped. The 1$^{st}$ stage safety valve is normally set at 8 bar gauge and the 2$^{nd}$ stage 5% higher than the final design pressure. The safety valves are normally set by the manufacturer and they may be sealed by the compressor maker after being tested and fitted on the machine.

If fitted at the machine or remotely, oil and moisture separators are classified as pressure vessels and a safety valve is required to be fitted and set at the final stage design pressure.

## 3.18 Compressor Cooling System

Air compressors can be either air or water cooled. The latter may use fresh or salt water.

### *3.18.1 Direct Seawater Cooling*

The majority of manufacturers who supply seawater cooled compressors presently use zinc or equivalent sacrificial anodes to protect the jackets against the corrosive effects of salt water. These anodes or rods are fitted in both the cylinder and cooler housings and the rods need to be checked regularly for wastage about every 1000 running hours or twelve months. The amount of wastage on each anode depends on the salt content. It will also be necessary – although a little more difficult – to check the compressor sea circulating system water piping for corrosion on a regular basis.

If anodes are fitted in a fresh water cooled machine a film will develop over the rods. If the machine is later changed to seawater cooling the film will prevent the rod from corrosion resulting in attack of the compressor material. In this instance new anodes should be installed.

In general pipework water velocities should be kept to a maximum of about 2.5 m/s although in short lengths of piping a maximum of 3 m/s may be allowed.

If a water pump is supplied either a direct drive or a vee belt unit driven from the compressor crankshaft may be supplied. Fig 18 shows a typical motor driven water pump. An alternative is a separate motor driven pump with safeguards to ensure that the pump will start and stop at the same time as the compressor. This is important because if cold water continues to flow through a hot stationary machine condensate will form on the cylinder walls entering the sump and cause contamination of the lubricating oil.

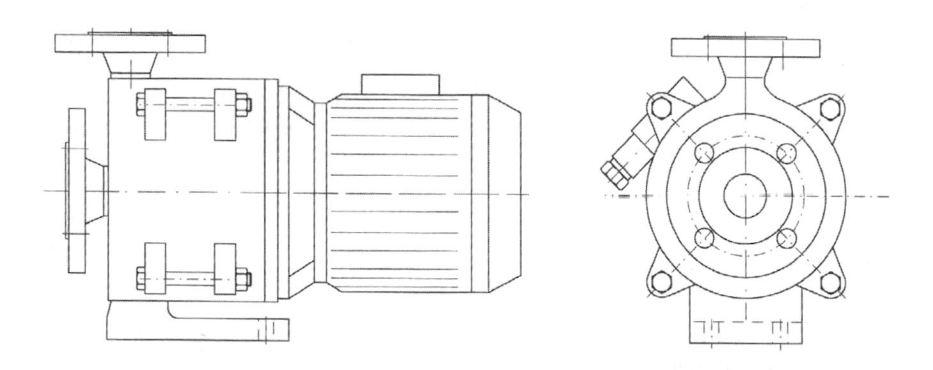

Fig 18: Motor Driven Water Pump

With low seawater temperatures around 4°C the compressor air discharge temperature may be lower than the initial temperature which will more than offset the radiated heat loss. Fig 19 shows a typical seawater cooling system.

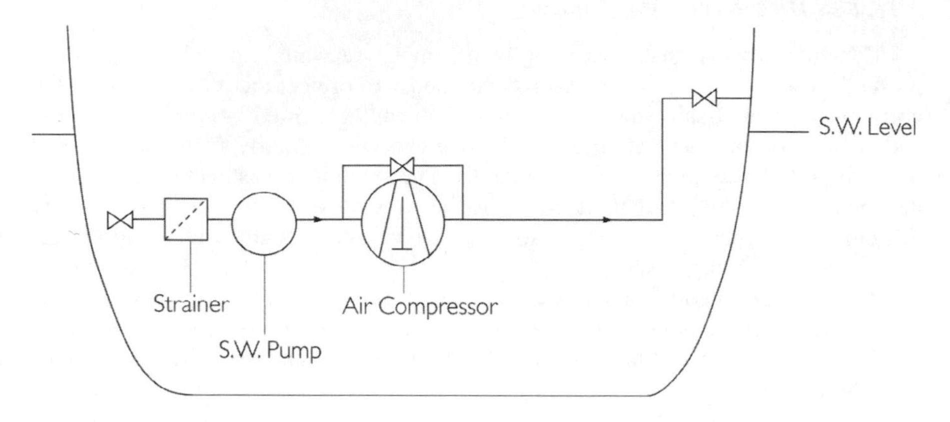


Fig 19: Seawater Cooling System

### *3.18.2 Direct Fresh Water Cooling*

The use of fresh water as the cooling medium is again more popular. The first cost of the installation is higher than direct seawater cooling due to the supply of installing a closed circuit fresh water system consisting of additional water pumps, heat exchangers, header tanks and additional piping and fittings. The shipbuilder in consultation with the shipowner decides if the cooling water will come from a central cooling water heat exchanger or from a heat exchanger that is dedicated to compressors only.

The velocities of fresh water pipework can be accepted up to 3.5 m/s without problems. The fresh water cooling temperature should be kept to as low as possible ie below 38°C and as recommended by the survey authorities.

In the following calculation for a water cooled compressor the power absorbed by the compressor is used since the bulk of the heat that is rejected to the cooling water will come from the pistons and piston rings. The heat from the main bearings, big end bearings, etc will not be transferred to the cooling water, but radiated away.

Heat to be removed by heat exchanger equals maximum power absorbed x 10.7kcal/min x maximum fresh water cooling flow.

Assuming
(1) Maximum seawater temperature of 32°C
(2) Fresh water temperature at cooler outlet not to exceed 38°C

$$\text{Fresh water temperature rise in °C} = \frac{\text{Maximum power (kW) absorbed x 10.7}}{\text{Fresh water flowrate (l/min)}}$$

## 3.19 Piping

It is often assumed that one main start air compressor will be running at a time and the pipework would be sized for the air flow of only one machine. This is incorrect since there is the possibility of two or more units running if the air demand so requires. The pipework should be sized to cater for the flow of all of the main start air units as if they were running simultaneously.

Fig 20 shows a badly designed seawater cooling system, which is often installed in ships. If the compressors are running to charge up the main air receivers, then the main engine is started and will draw the majority of the cooling water through the main engine coolers, so depriving the compressors of cooling water. This will result in the compressor having a piston seizure. If the compressor driven water pump is of the positive displacement type, it will act as an automatic shut off valve on a stopped machine. If the cooling water is taken directly from a central cooling system it will be necessary to fit an automatic shut off valve.

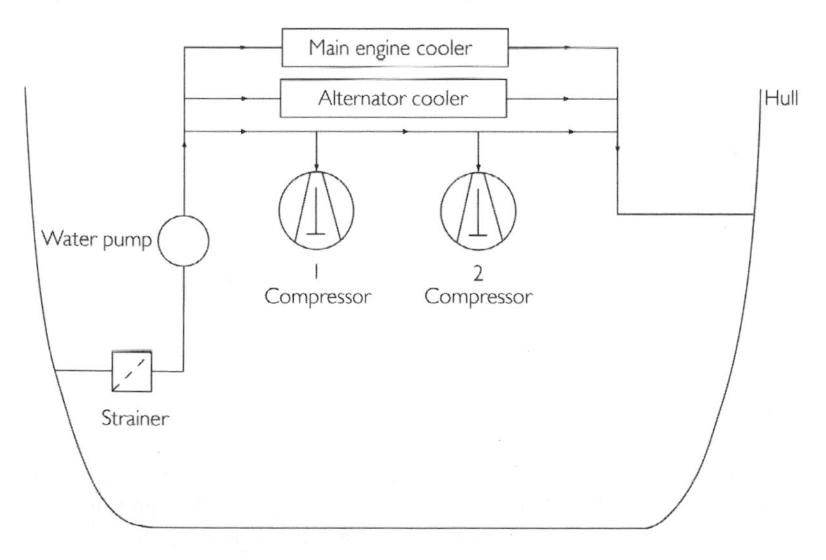


Fig 20: Badly Designed Cooling Water System

## 3.20 Lubricating Oil System

The lubricating oil system of a reciprocating compressor can be either splash or pressure lubricated.

In small air compressors lubrication is normally by splash and splash rods will be fitted to one of the connecting rods, see Fig 21. The turning of the connecting rod will cause the rod to dip into the oil in the sump and create a splash effect to lubricate the main bearings etc. An oil mist is created, which will provide for lubrication of the small end bearings and cylinders. Optional automatic machine shut down on low oil level or low oil pressure, may be fitted.

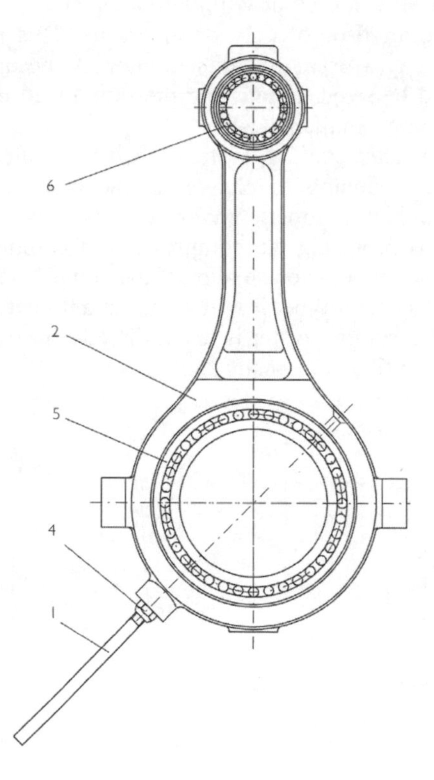


1 – Splash rod x1
2 – Connecting rod – 1st stage x1
4 – Hexagon nut x1
5 – Connecting rod bearing x1
6 – Piston pin bearing x1

Fig 21: Connecting Rod with Splash Rod

Forced pressure lubrication is preferred on compressors of over 80 m$^3$/h output. The crankcase oil pump draws oil from the sump via a suction filter and delivers oil under pressure to the bearings. Remaining areas are lubricated by the resultant oil mist.

A recent but successful innovation is for the crankcase oil breather to be vented direct to the 1st stage suction, resulting in a pressure of below 0.4 bar in the crankcase. This method also helps to lubricate the first stage valve. Not having an open breather to the atmosphere prevents oil covering the compressor and surrounding areas and avoids engine personnel breathing toxic oily compressor fumes.

Fig 22 shows a modern oil pump fitted to the compressor driven from the crankshaft. It is important that the oil losses are kept to a minimum especially at the time of the initial start-up when the oil will be cold and viscous.

For accessibility for maintenance purposes the oil pump should be fitted on the outside of the compressor.

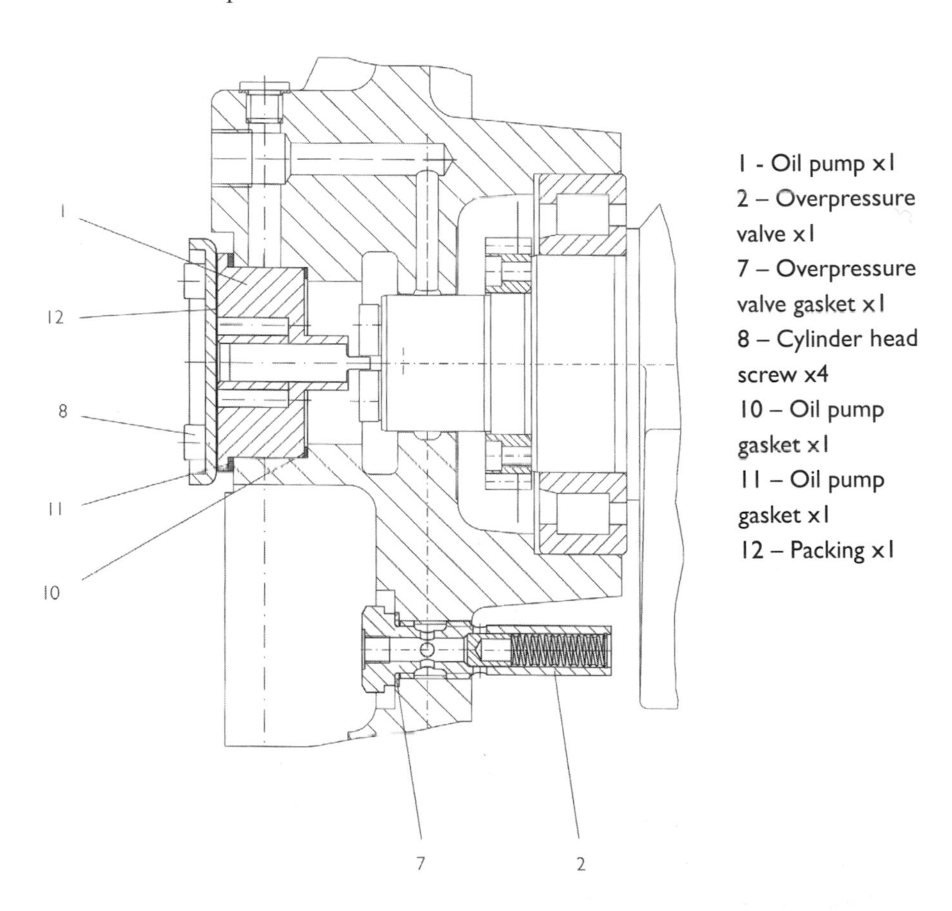


Fig 22: Oil Pump

The amount of heat created within the crankcase will be relatively low and the fitting of a special heat exchanger to reduce heat from this source is not a necessity.

A low oil pressure cut out switch is required for larger compressors.

To prevent damage to the oil pump transmission at initial compressor start with cold, high viscosity oil, it is necessary to fit a relief valve in the pressure side of the oil system. This valve should be sized to prevent an overpressure of 25% above normal working oil pressure.

## 3.21 Air Suction Filter/Silencer

Although there are many different types of air suction filters/silencers, it is unusual to find either the cell type or the oil bath type in service onboard. It is possible that the dry filter type can be found in service in the marine field. This filter has a filtering media of either felt cloth or paper, the latter being slightly oil coated.

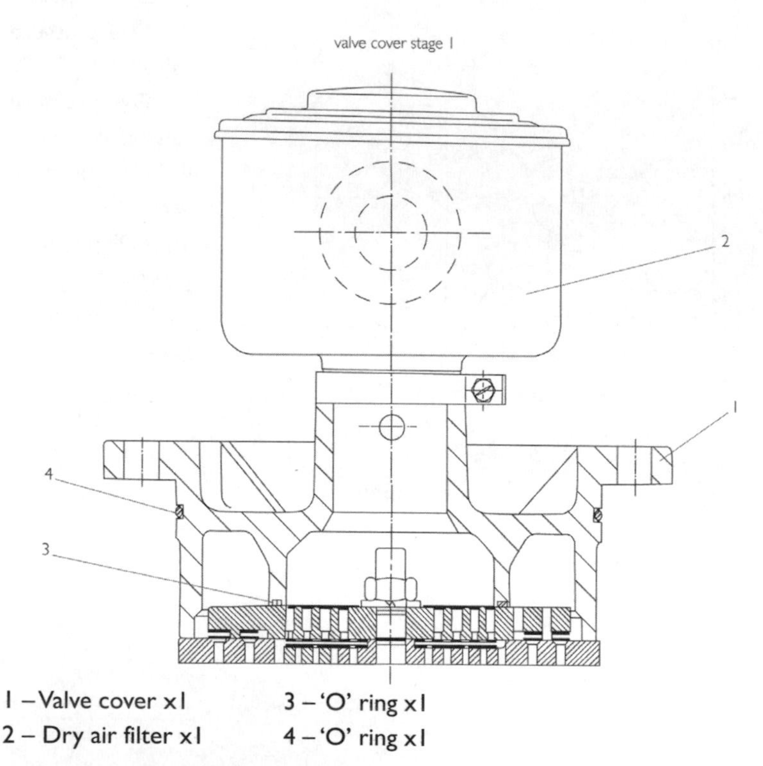


Fig 23: Suction Filter/Silencer

The filter most widely used in marine compressors is of the cylindrical design. The elements can be removed and cleaned with a solvent then blown dry with low pressure air before replacing the clean element. The oil coating on the element traps dirt although the engine room atmosphere drawn into the compressor intake is normally fairly clean.

It is important that the air filter is cleaned according to the manufacturer's instructions so that the pressure drop across the filter is minimal. Another important recommendation is that no fumes with flammable or explosive properties eg, paint are drawn into the compressor while running. The temperature of the air can be raised to a dangerous level (above oil flash point and close to ignition point), due to excessive restriction caused by a dirty or damaged air filter. Fig 23 shows a typical suction/filter silencer.

If very clean air is required it may be desirable to draw air from a clean room area. In such a case it is necessary to fit a filter/silencer which has a flange or equal fitted so that pipe ducting is led to the compressor. The compressor inlet would need a similar flange or equal.

## 3.22 Oil and Moisture Separators

It is recommended that oil and moisture separators are fitted in the air system. The most effective method is to fit oil and moisture separators after both the final intermediate and final stage of compression to ensure that a large percentage of the oil and water content in the air is removed before the air is passed downstream. It is preferred that these separators are fitted and piped up on the compressor itself to save space and reduce installation costs. By fitting a timer into the control panel the oil and moisture separators can be arranged to blowdown automatically at pre determined intervals, ie, every 15 to 20 minutes dependant on the operational conditions eg, areas of high humidity.

If the compressor has three stages a separator should be installed after the final two stages or for a four stage unit, after the final three stages of compression. As the separators are pressure vessels they have to be tested and stamped by the classification surveyor and it is a requirement to fit a relief valve on each separator.

The alternative to mounting the separator on the compressor is to install them downstream of the unit itself and one separator for each compressor should be installed. If the compressor units are installed close together it may be possible that only one separator sized for two or even three compressors can be installed. Again the separators should be arranged to blow down automatically and it is standard practice for manual drains to be fitted as a back up.

Several different types of separator can be utilised:

### 3.22.1 *a) Vortex type*

This type of oil and moisture separator unit is probably the most commonly used for compressors. Air entering the separator is forced into a vortex movement so that centrifugal forces throws the free moisture clear and to the side of the separator where it is trapped and drained to an oil and moisture collecting tank but in some installations direct to the vessels oily bilge. The separator should be fitted with an automatic but at least a manually operated drain. A typical design of a vortex separator can be seen in Fig 24.

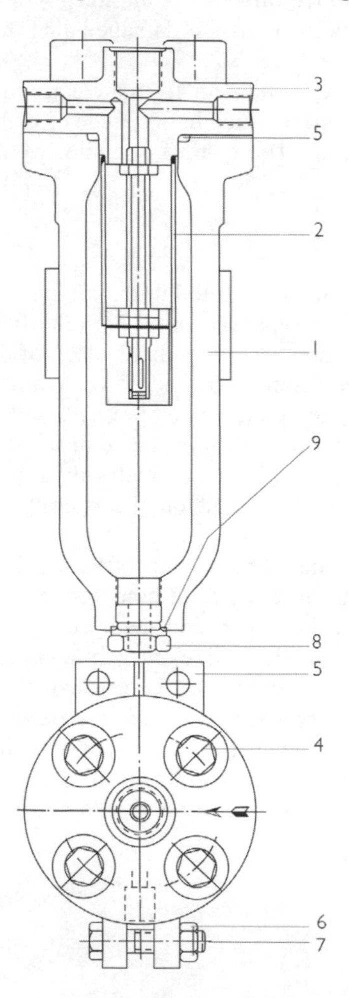


1 – Separator body x1
2 – Filter insert x1
3 – Separator cover x1
4 – Filter head screw x4
5 – Mounting x1
6 – Nut x1
7 – Bolt x1
8 – Drain plug x1
9 – Drain Seal x1

Fig 24: Vortex Separator

### *3.22.2 b) Directional Change Separator*

This type of separator works by suddenly changing the direction of the air at least once through 180deg but sometimes more. This separator also reduces the upward velocity of the air to a very low speed so that the moisture droplets are carried to the bottom of the vessel due to the directional change. Again drainage should be carried out automatically rather than manually.

### *3.22.3 c) Coalescing Type*

This type of moisture separator functions at a higher efficiency than above but requires more attention as it tends to foul up more rapidly. There are two types: one is of the mesh type which coagulates the moisture on the strands and the other is of the porous ceramic type unit in which the moisture coagulates around its pores. This latter is more efficient but should be preceded by a coarse type filter.

## 3.23 Drain Collecting Tank

Fluids draining from the oil and moisture separators can either be led directly to the oily bilge or to a collecting tank. The latter can either be supplied as an integral part of the compressor unit or installed nearby. The tank can be designed to reduce the noise level to below 85 dB(A) at one metre distance when the compressor drains blow down. If the tank is of the 'closed type' it will be necessary to fit a relief valve.

## 3.24 Oil Content in Compressed Air

The oil carried over into the air system by the compressor can be considered in two topics:

1) The risk of fire and explosion in the oil lines within the compressor and downstream to the air receiver; this is particularly relevant to oil pressure lubricated compressors.

2) Removal of oil from compressor discharged air, as in the case of a lubricated oil type of compressor being used for control air, or, if the air was being used for breathing air purposes.

Examining the above in greater detail:

1) Before 1970 there were many discussions in technical journals by the manufacturers of air and gas compressors regarding the potential risk of fires and explosions within the air piping of oil lubricated reciprocating compressors. Recommendations on how to overcome these serious problems within basic compressor and cooler design were given but achieved limited success.

Following these design ideals today has seen the introduction by many manufacturers of 90deg 'V' configuration, reciprocating machines. These are now manufactured using superior machining techniques leading to improved component accuracy. Installation of oil and moisture separators with automatic drainage and located at least after the final stage of compression significantly reduce the amount of oil and moisture carried downstream.

Both lubricated and non lubricated compressors would have practically the same safety level in operation. An oil lubricated machine fitted with an oil and moisture separator would expect to have an oil carry over of no more than 7 $mg/cm^3$ into the air system. If an oil removal filter was also fitted directly after the oil and moisture separator the oil carry-over into the air system should be further reduced to no more than $1 mg/cm^3$.

Even if an oil free compressor was installed in the engine room for main start air duties the suction air would be oil laden and oil content of above $1 mg/cm^3$ would still be expected to be present in the delivered air. It would be essential to fit oil removal filters after the compressor air discharge to meet specification requirements.

2) As mentioned above and depending on the amount of atmospheric pollution the compressed air delivered from an oil free compressor has an oil content that cannot be neglected. The type of piston rings used for oil free compressors have a carbon coated surface which cause dust in the discharge air and a debris collecting filter would also have to be installed.

If air is used for breathing air purposes it will be very important to remove any oil vapour to meet the international breathing air standards (ie, BS4275, DIN 3188), and an adsorption filter should be additionally fitted. Depending on the capacity many breathing air filters have all the necessary requirements in one filter body. If large flows of breathing air are required more than one filter housing may be necessary.

## 3.25 Supervision

In order that compressors can be supervised locally the gauges and protection devices that should be fitted to each unit are indicated in the accompanying Table. The switches give a signal only and in general these are used for alarm and stop functions.

The final compressed air outlet temperature switch is for the supervision of the cooling water flow. On cooling water failure the switch will activate on reaching the set value (eg, 80°C), and shut down the compressor.

| Pressure gauge | 1st stage air pressure |
| --- | --- |
| | 2nd stage air pressure |
| | 3rd stage air pressure (if three stage machine) |
| | Oil pressure |
| Thermometer | Cooling water inlet (if water cooled) |
| | 1st stage compression temperature before cooler |
| | 2nd stage compression temperature before cooler |
| Mechanical safety device | Safety valve 1st stage |
| | Safety valve 2nd stage |
| | Safety valve 3rd stage (if three stage machine) |
| | Rupture plate for cooling water pressure (if water cooled) |
| Electrical switches | Oil pressure switch (P < 1 bar) |
| | Final compressed air outlet temperature (T > 80°C) |
| | Cooling water shut off valve |

Gauges and protection devices that should be fitted to each compressor

## 3.26 'B' Screw Compressors

### *3.26.1 Screw Compressor Background*

Air cooled screw compressors have been used extensively in industry for factory air applications since the late 1970s, taking over from the vertical and 90deg 'V' models of reciprocating compressor types previously used. It was not until the early 1980s that saw the introduction of screw compressors for control air and general service air applications onboard in preference to reciprocating compressors. These screw compressors have proved to be more reliable and easier to maintain than the reciprocating compressors.

With outputs of up to several thousand $m^3/h$ and operating at pressures of up to 13 bar silenced (noise levels below 75dB(A) at one metre), air cooled screw compressors have now become the ideal selection for general service and control air duties onboard. Fresh or seawater cooled machines are also available but because of additional water pumps, heat exchangers, fittings and pipework within the units, the space envelope and weight will be much greater. For control air purposes oil free screw units, both air and water cooled are available.

A typical oil lubricated screw compressor profile is shown in Fig 25 but individual manufacturers have slight differences in their design.

Fig 25: Typical Oil Lubricated Screw Compressor

## 3.27 Suction Filter

A completely load free optimum start up is provided by the fitting of a closed suction regulator with metered bypass to ensure that that there is no vacuum created during start-up and idle operation. This will eliminate start up peaks and reduce energy consumption.

The air drawn into the compressor is atmospheric air normally drawn directly from the engine room or the room where the unit is installed. Alternatively the air can be drawn from a clean air area via air trunking. The inlet air filter will separate solid impurities such as dust particles from the intake air. Removal of such particles will help to reduce wear on compressor rotors and assist in providing the system with clean air.

The compressor rotors are normally manufactured from G22 treated steel and the two rotors are connected by synchronised transmission so that the surface profiles do not touch.

The suction filter should be easily accessible for filter replacement.

## 3.28 Screw Profile

Screw compressors operate on the displacement principle having two parallel rotors with different profiles working in opposing directions inside the screw housing. The intake air is compressed to its final delivered air pressure in chambers which reduce in size through the rotation of the screw rotors. Fig 26 shows the compression process of the screw compressor.

Phase 1- The air is drawn through the suction entrance and then into the open screw profile of the rotors on the inlet side.

Phase 2 and 3 - The air inlet will close by the continuous rotation of the rotors. The volume of air will reduce and the pressure will increase. Oil is then injected onto the screw rotors at this stage.

Phase 4 - The compression process will be completed and the final pressure reached and the discharge of the air will begin.

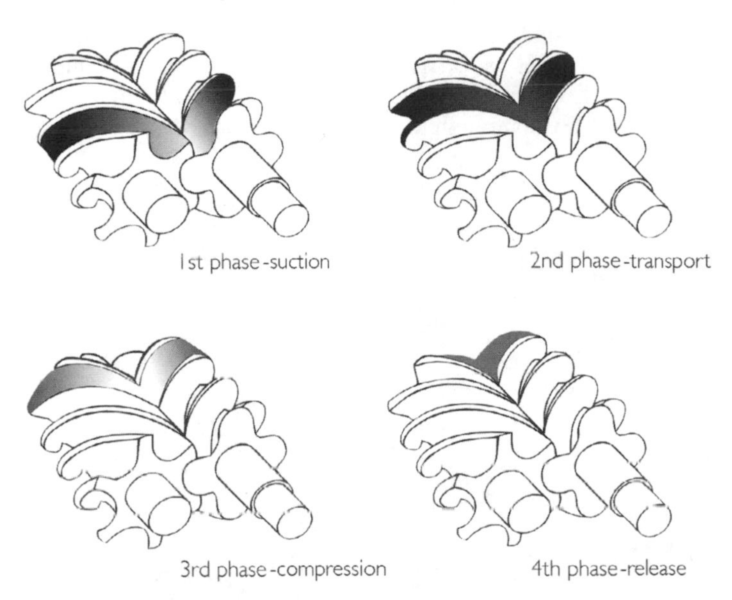


Fig 26: Screw Compressor Process

## 3.29 Air Path

The electric motor drives the compressor by the use of vee belts. These belts are designed for a life of approximately 25 000 hours.

Intake air flows from the suction filter to the suction valve and into the compressor stage where the air is compressed. The relief valve closes and the suction valve will open at the direct-on-line switch point and the reverse way at the upper pressure point and into the ideal running phase. The oil separator decreases the residual oil to between 2 and 4 mg/cm$^3$. In the compressed air recooler the compressed air temperature reduces to about 10-13°C above the ambient and leaves the machine through the compressed air connection. Fig 27 shows diagrammatic air flow system.

## 3.30 Oil Circuit

Cooling and sealing the lubricating oil flows into the compressor stage by the pressure that is produced in the oil tank. The oil leaves the compressor stage mixed with the compressed air and according to regulations a safety valve must be installed to protect the machine against excess pressure. The oil is separated from the air by up to 98% utilising the design of the tank which collects the oil. The residual separation of the oil is carried out by the oil separator and the oil temperature is optimised by the temperature controller. The oil enters the bypass line or the oil cooler and flows through the oil filter into the compressor stage.

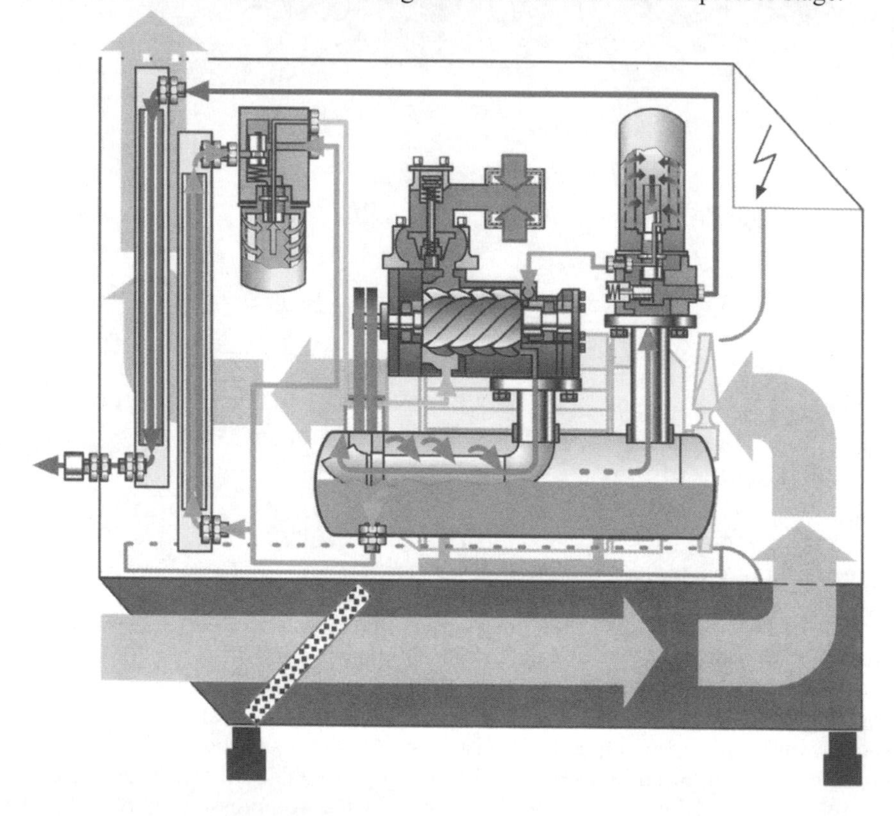

Fig 27: Diagrammatic Air Flow System

### *3.30.1 Oil Separator and Collector*

A horizontally mounted oil separator/collector fitted with an oil drain cock is included within the package. The unit will ensure a high pre-separation rate of up to 98% of low oil foam production via its large oil surface. Low residual oil content in the compressed air, long service life of the oil separation cartridges and an oil drain cock ensure a clean oil change without the need for special equipment.

### *3.30.2 Oil Temperature Control*

The oil will quickly reach the ideal working temperature and is controlled automatically. Cooling is effected by a finned cooler with a large surface area that can be cleaned easily. The marine version is designed for operation at ambient temperatures of up to 50°C.

## 3.31 Electric Motors

The screw air compressor is normally driven by drive belt transmission through a totally enclosed fan cooled electric AC motor which operate at speeds of up to 3000 rev/min. The motors will be located on a sliding plate for correct belt tensioning.

## 3.32 Electronic Control

This is a plain language display monitoring all operating states of the compressor. Actuating variables should be included in its memory controlling all the functions and automatic selection of the economic mode of operation.

Extensive self protection, early warning features, servicing reminders and fault diagnosis are some of the major features that are now included in the screw compressor unit memory. A more detailed description of the automatic monitoring of screw compressors can be found in Chapter 5.

## 3.33 Sound Insulation

A screw compressor without sound insulation panels will have a noise level of over 90 dB(A) at one metre distance which is unacceptable. Today's machinery spaces have a maximum noise level target of a 85 dB(A) at one metre. By using sound damping panels the noise levels of a screw compressor can be reduced to below 75 dB(A) at one metre distance.

In addition the compressor unit assembly can be isolated from the base frame of the unit by fitting resilient marine type mounts.

## 3.34 Principle of Operation

Described here is a typical operation of a screw compressor. The atmospheric air from the machinery space is drawn into the compressor via the controller and cleaned by an air filter. Then in four phases the air is compressed to the final required pressure by two screw rotors. The air end is driven by an electric motor. During operation a continuous flow of oil is flushed through the compression chamber for cooling, lubrication, and sealing purposes.

Later in the first and second stages the compressed air and the oil are mechanically separated with an efficiency of around 98% and then transferred to an oil separating cartridge where a separating efficiency of almost 100% is achieved. Only a small residual oil content of 2-4 $mg/cm^3$ will be retained in the compressed air. The hot oil is fed through an oil cooling unit and oil filter and then clean oil is injected back into the stage.

## 3.35 Flexible Hoses

If the compressor unit is mounted on resilient mounts the air outlet should have a flexible hose fitted between the compressor and the ship's piping so that no forces can be transmitted. The flexible hose can be supplied by manufacturer or shipyard.

## 3.36 Non-Return Valve

A non-return valve installed between the piping system and the air receiver is not required as this valve will be included within the screw air compressor package. It is strongly recommended to install a shut off valve in the air discharge line so that the compressor can be isolated from the air piping system to safely facilitate maintenance or repair.

## 3.37 Drain Trap

Within the air system but after the compressor automatic condensate drain traps are recommended to be fitted to collect any condensate present in the pipework. If an oil and moisture separator is fitted after the compressor the presence of any oil in the drain traps is considerably reduced.

An electronic alarm/monitoring system can be installed after the air compressor to ensure permanent monitoring of the operational state of the condensate drain. Any malfunction will be signalled via the potential free alarm contact.

## 3.38 Screw Compressor Materials

The manufacturers of screw compressors will have their own material specifications but a typical material list is shown in the Table.

| Frame | Rimmed steel, St 37 |
|---|---|
| Compressor stage casing | Grey Cast Iron, GG20 |
| Rotors | Treated Steel, G22 |
| Oil tank | Rimmed Steel, St 37 |
| Filter boxes | Aluminium |
| Air coolers | Aluminium |
| Fan Wheel | Plastic |

## 3.39 Compressor Air Receiver

In the majority of cases a separate air receiver would be supplied for either control or general service air but for small screw compressors with a delivered air capability below 100m$^3$/hr it is possible to mount the compressor on top of a horizontal air receiver completely piped and wired up.

The advantages of such a unit would be saving in space and lower installation costs. Fig 28 shows a typical screw air compressor mounted on a horizontal air receiver.

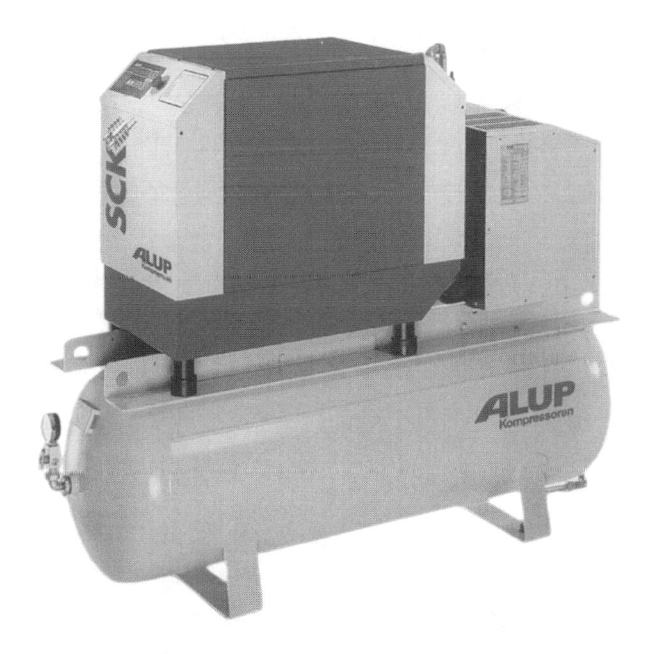


Fig 28: Screw Compressor Mounted on Receiver

## 3.40 Oil Free Screw Compressor

The majority of screw compressor manufacturers will include an oil free screw compressor within their range for control air duties.

An alternative would be to supply a range of filters to remove oil content in the air. Most oil removal filters will remove 99.99% of oil from the air.

## 3.41 Screw Compressor Cooling

### *3.41.1 Air Cooling*

The majority of compressors supplied for general service or control air applications are of the air cooled type.

### *3.41.2 Water Cooling*

If required seawater or fresh water units may be supplied but these machines occupy more space and are heavier than an air cooled compressor and hence have a considerable cost disadvantage.

## 3.42 Conclusion

The table listed below gives a summary of the type of compressed air application and the type of air compressor which may be employed:

| Application | Compressor Type | Water or Air Cooled |
|---|---|---|
| | | |
| Main Start Air | Reciprocating | Water, Air |
| Emergency Air | Reciprocating | Air |
| Working Air | Screw or Reciprocating | Air, Water |
| Control Air | Screw or Reciprocating | Air, Water |

# Chapter 4

# INSTALLATION

## 4.0 Siting

The siting of a marine compressor onboard whether installed in the engine room or on an upper deck level is not usually critical. The air system pipe runs are relatively short and it is unnecessary to site the compressor units at a mid system point for either a reciprocating or a screw unit. Care must be taken to ensure that the compressor unit is positioned so that it draws clean atmospheric air via the suction filter/silencer, clean atmospheric air, unpolluted by oil, steam or gas leakage. Even the picking up of hot air can be detrimental to the performance of the compressor in that it will reduce the output and increase the interstage temperatures consequently increasing maintenance levels.

On larger units it is important that facilities for attaching lifting equipment above the compressor and prime mover are available. Lifting should be carried out in accordance with the manufacturer's advice and should be contained in the instruction book. Fig 29 shows a typical compressor set lifting arrangement.

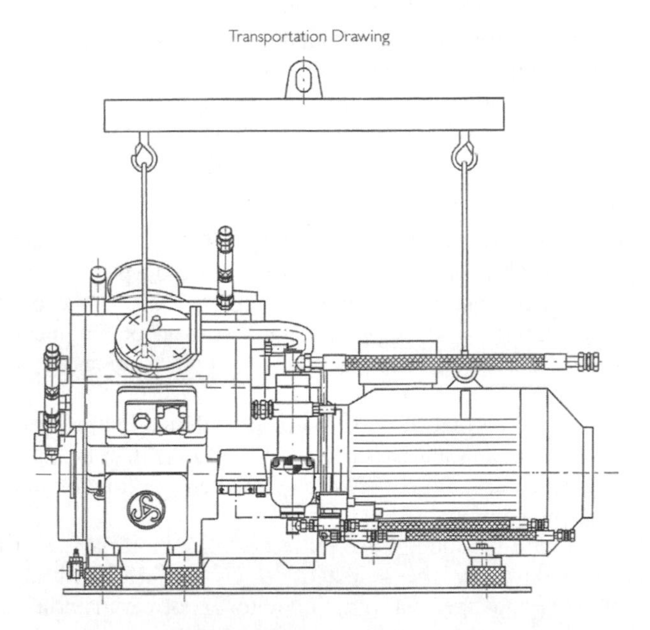


Fig 29: Compressor Lifting Arrangement

## 4.1 Compressor Installation

For machines with cylinders arranged in a vertical configuration and the prime mover coupled to the compressor via a flexible coupling the unit is mounted on a bedplate. It is standard practice for the unit to be mounted on 'Teco' pads or similar. The pads would be fitted between the compressor unit and the engine room seatings. The pads will help to absorb some of the downward forces created by the vertically arranged pistons.

Recent developments have seen the introduction of new technology, with the compressor pistons arranged in either a 90deg 'V' or a 'W' configuration which has resulted in a reduction of vertical structure borne forces. In addition the prime mover has now been direct (flanged) coupled to the compressor with the introduction of a bell housing resulting in a rigid structure. The fitting of resilient mounts under both compressor and prime mover has meant that the base frame is no longer required making savings of weight, capital and installation costs. It will be necessary to install a subframe (but not a heavy bedplate), if additional equipment such as a control panel is included in the package. Fig 9 in Chapter 2 shows a typical 90deg 'V' resiliently mounted water cooled unit.

It is also possible for this type of resilient mounting arrangement to be arranged on a vertical compressor mounted on a bedplate.

The screw compressor unit will normally be fitted with a silenced canopy and the complete package will be mounted on a subframe with resilient mounts which are fitted between compressor frame and engine room seatings. Due to the rotary screw configuration the out of balance forces of such a unit are very low.

## 4.2 Pipework

The majority of compressors installed are fitted with oil and moisture separators, fitted and piped up on the unit itself, rather than have the pipes supplied loose and separators installed downstream of the compressor unit. Some of the older air systems are not supplied with separators and it is good engineering practice that at the lowest point of any pipe bends and runs, condensate traps (preferably of the automatic type), are fitted.

Air pipes should be arranged so that they have a slight inclination away from the compressor to ensure that condensate build up is not allowed to find its way back into the compressor. Again an automatic type of condensate trap should be fitted at the lowest point of the pipe run.

All pipes and fittings should be adequate for the design air pressure and be according to the rules laid down by the selected design authority. Air above atmospheric pressure can be very dangerous to the operator if correct materials are not selected.

## 4.3 Air Cooled Compressors

When designing the compressed air starting system for a ship almost the first phrase written down in the compressor specification is 'of the water cooled type'. It should not be forgotten that the greater majority of small coastal vessels, offshore supply boats and deep sea tugs are started by two stage, air cooled compressors which do not suffer excessive maintenance problems. Presently an increasing number of vessels, from cruise liners to cargo ships are equipped with air cooled starting air, control air and general service compressors, and several thousand of such units are in operation worldwide.

With the introduction of three stage air cooled compressors operating at pressures of up to 40 bar, the interstage air temperatures are claimed to be lower than a two-stage water cooled machine. Outputs of 400 $m^3/h$ and over are achieved by using multi stage air cooled units. It is claimed that these machines have a much better reliability than water cooled compressors of similar capacity. This is supported by the fact that a two year guarantee is offered as standard in place of the traditional one year guarantee offered for two stage water cooled machines.

It must also be remembered that both the two stage and three stage units will need a cool supply of air to be drawn across the machine [as shown in Fig 8]. For more detailed installation information reference should be made to the manufacturer's instructions.

It is also recommended that the ambient air temperature in the engine room remains in the range 5°C to 48°C.

Due to the absence of water cooled jackets and intercoolers air cooled machines do not suffer unduly from cooler fouling or corrosion. An air cooled unit does not have the maintenance load of water piping, valves and fittings. Three stage units have lower interstage temperatures so there is less carbon deposition on valves and cooler tubes.

A typical comparison of air and water cooled compressor interstage air temperatures is listed below with a compressor operating at a pressure of 30 bar and a speed of 1770 rev/min.

| Description | Air Cooled | Water Cooled |
|---|---|---|
| 1st stage | 140°C | 125°C |
| 2nd stage | 140°C | 140°C |
| 3rd stage | 155°C | n/a |

The above is based upon a water temperature of 32°C and an air temperature of 48°C.

## 4.4 Water Cooled Compressors

With the introduction of the 90deg 'V' and 'W' configuration compressors there is a marked reduction in out of balance forces in the machine. This is one of the factors in improved reliability of such compressors compared with older vertically designed units.

Fitting rubber vibration pads and the absence of a base plate has resulted in compressor movement on the mounts. This does not cause problems with the compressor but the fitting of flexible pipes between the machine and the ship's air, drain and water systems are required.

The noise level of a water cooled compressor is slightly lower than an air cooled unit because the water jackets have a sound damping effect.

For details of the various drive arrangements see sections 4.5.1, 4.5.2 and 4.5.3. Cooling water can either be obtained direct from the sea or by using a fresh water closed circuit heat exchanger.

## 4.5 Compressor Seatings

The air compressor and its prime mover will be vee belt driven, direct coupled via a flexible coupling, or flange coupled.

For a vertical reciprocating compressor it is standard practice for the unit to be installed on an adequately stiffened bedplate to assist in the reduction of the vibrations. Such vibrations are detrimental to compressor components. The vertical compressor has high out of balance forces caused by the inertia masses. This makes it important that the loaded area of the compressor is spread over as wide an area as possible so that vibrations and out of balanced forces can be absorbed by the structure of the base frame.

Fig 30 shows a typical vertical compressor foundation and the dimensions of A, B, C and D will depend upon the capacity of the compressor. Considerations of various available drive options follow.

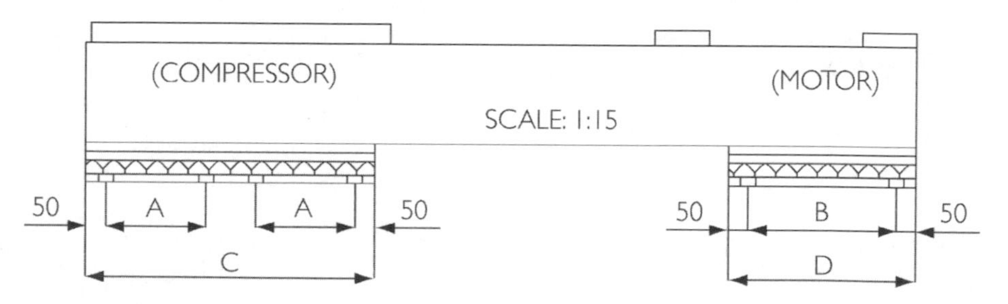


Fig 30: Compressor Foundation

### *4.5.1 Vee Belt Drive*

If a compressor is vee belt driven from its prime mover via a flexible coupling, it is necessary to mount the complete unit on a bedplate to create a rigid structure. This will prevent any movement between the compressor and its prime mover which may be an electric motor or diesel engine. It is necessary to fit 'shims' under both the compressor and the prime mover to ensure the complete unit is level. The levelness should be checked both when it is first installed and after any major maintenance work ie, if the compressor or motor are removed.

The bedplate will be of a substantial design to take the weight of the compressor unit plus the weight of any additional fittings eg, a water pump or compressor control/starter panel. This is more important when several compressor units or air receivers are installed on the same bedplate.

### *4.5.2 Direct Coupled via Flexible coupling*

It is necessary for the complete unit to be mounted on a bedplate and the comments in the previous section are equally applicable. The compressor unit may be of the vertical or 'V' type, (not fitted with bell housings). Fig 31 shows a typical flexible coupling.

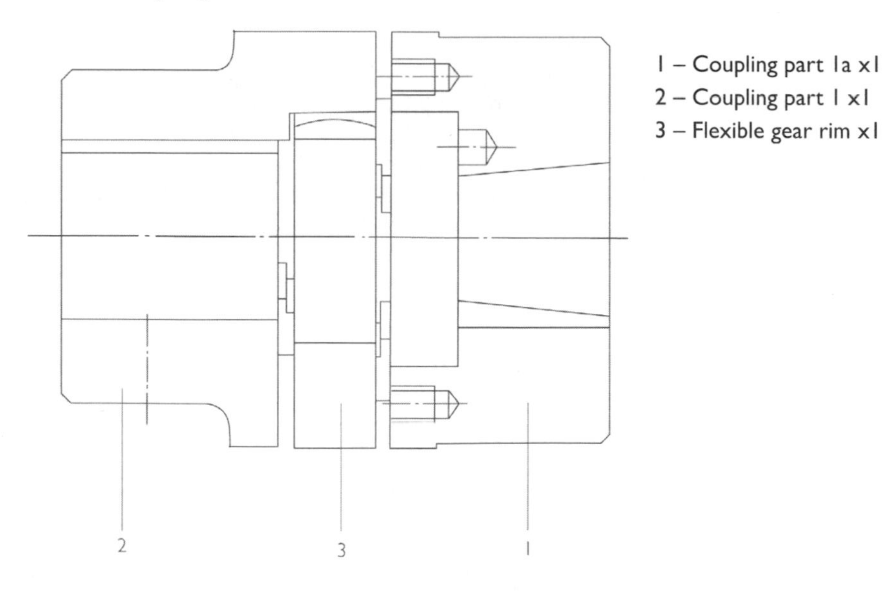


Fig 31: Flexible Coupling

### *4.5.3 Flange Coupled*

The majority of modern designs of air and water cooled compressors have the compressor flange coupled by a bell housing to the prime mover whether it is an electric motor or diesel engine. This configuration will enable a three or four point mounting system to be utilised doing away with heavy and expensive bedplates and the need to fit shims. Furthermore the compressor set will have a smaller space envelope, a lower unit weight and reduction in maintenance costs due to the absence of guards, etc, that would otherwise require removal.

## 4.6 Flexibles

Modern 90deg ‘V’ type compressors are normally fitted with rubber based resilient mounts which are placed below the compressor unit feet to absorb both the structure borne noise and movement of the vessel. In some installations where bedplates are still fitted, solid pads are placed at strategic positions between the compressor set and the bedplate. The calculation for pad location is the responsibility of the compressor manufacturer.

When flexible mounts are fitted, flexible pipes have to be installed at any point where a pipe on the compressor is connected to the ship’s piping system ie, water inlet and outlet pipes, delivered air pipes, oil and moisture separators and main drain to bilge, etc. These flexibles can be supplied by either the compressor manufacturer or the shipyard. The correct and wrong methods of installation are shown in Fig 32.

Importantly the delivered air flexible pipes fitted to the ship’s air system have to be tested and approved according to the rules of the survey authorities. Flexible hoses require to be marked with their date of manufacture and their shelf life must be given to customers.

## 4.7 Prime Movers

Prime movers are normally electric motors designed for direct on line starting. For the emergency air compressor (also known as the black start unit), diesel engine drive is the most commonly used although electric drive is occasionally selected.

### *4.7.1 Electric Drive*

In recent years the specification of large electric motors have changed from enclosed ventilated drip proof (EVDP) motors to totally enclosed fan cooled (TEFC) motors with either IP54 or IP55 insulation being specified. The temperature rise to Class F is preferred to Class B. This means that electric motors are able to function in ambient air temperatures up to 60°C.

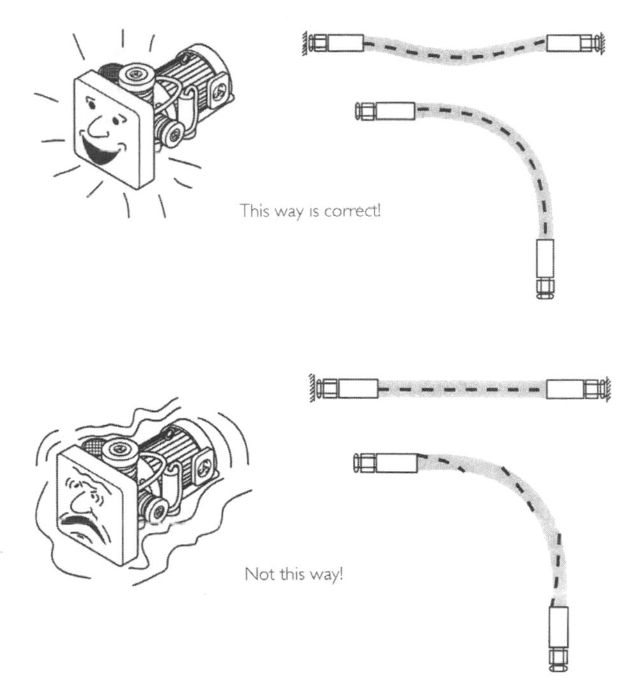


Fig 32: Flexible Pipes

For many years small TEFC motors up to approximately 15kW have been fitted to compressors mainly because the leading motor manufacturers had decided to discontinue manufacture of small EVDP motors for marine and industrial use.

Some specifications call for thermistor protection to be included within the motors. The embedded temperature sensors are only able to work in conjunction with a tripping device insofar as when reaching the limiting temperature, the thermistor changes its resistance almost instantaneously. This action is utilised in conjunction with the tripping device, to monitor the temperature. The relay which is incorporated in the device has a changeover contact in which the contacts can be used for the control system. The protection system is self monitoring and the low switching tolerance is approx 1kΩ to 3kΩ.

## *4.7.2 Diesel Drive*

The majority of small diesel engines used for starting emergency air compressors are started by using a manually operated starting handle. The diesel engine can either be direct driven or vee belt coupled to the compressor, with the direct coupling option more common to eliminate the complexity of belts, guards, etc. This option often results in elimination of a bedplate and utilises a four mount point support similar to the main start compressors. Fig 33 shows a typical diesel driven compressor set.

As an option to hand start and included as part of the package a 12V battery and charging alternator may be supplied.

## 4.8 Drive Shaft

If vee belts are utilised then drive shafts on both compressor and prime mover are fitted with grooved pulleys. The advantage is that it will enable the motor or diesel engine to run at a much higher speed ie, if the electric system is 60Hz the prime mover can run at 1760 rev/min. The disadvantage of this type of arrangement is that the complete unit is mounted on a bedplate and guards, etc, have to be fitted to ensure personnel do not touch any of the moving parts.

## 4.9 Packaging

Most compressor sets will comprise of a compressor, prime mover and possibly a fresh or sea water circulating pump, but with the control panel mounted on a nearby bulkhead or in the control room. To assist the shipbuilder the compressor supplier may offer two or three units that are packaged onto a single skid bedplate. This package may include compressors, prime movers, control panels, pipework, fittings and electric wiring. It is also possible to include air dryers and air receivers piped and wired up.

One advantage of packaging the equipment is that all the equipment is tested as a complete set and witnessed by the local surveyor or the shipbuilder before leaving the factory.

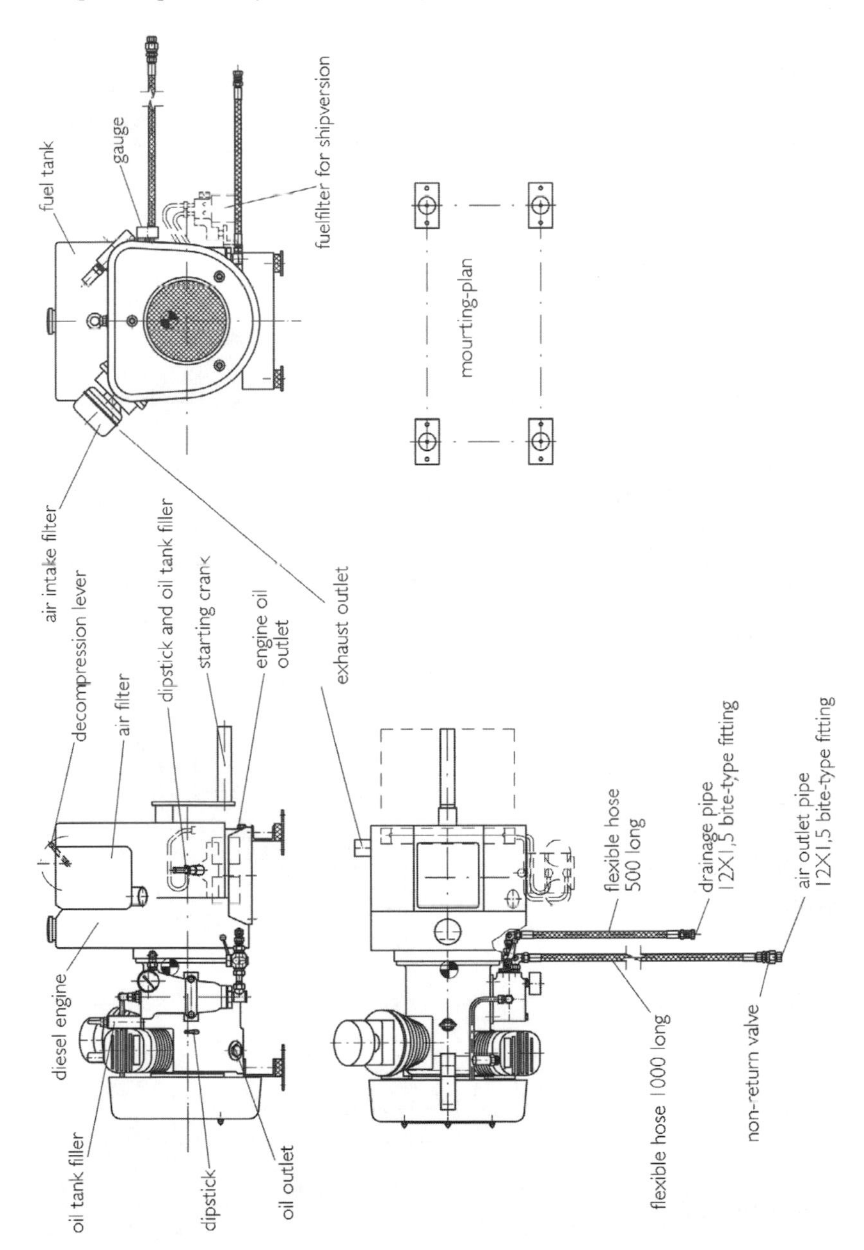


Fig 33: Diesel Driven Compressor

## 4.10 Air System

The design of the compressed air piping is critical and Fig 34 shows a typical three stage compressor schematic installation. The following fittings will be included within the system itself.

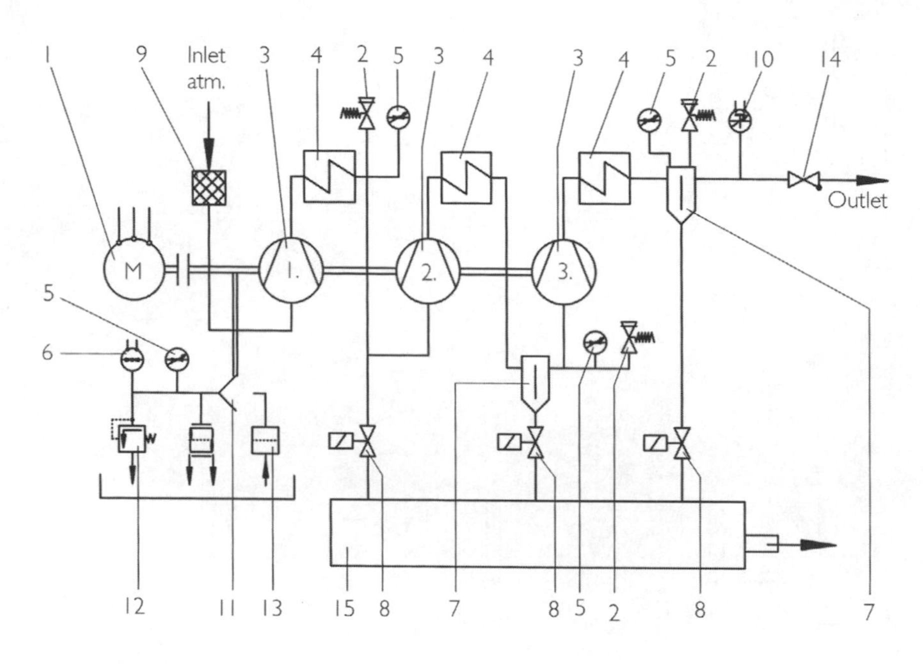


1 – Drive motor
2 – Safety valve
3 – Compressor stage
4 – Intercooler and aftercooler
5 – Pressure gauge
6 – Oil pressure switch
7 – Separator
8 – Solenoid valve
9 – Intake filter
10 – Temperature switch (optional)
11 – Oil pump
12 – Overpressure valve
13 – Oil filter
14 – Non-return valve
15 – Condensate collecting pot

Fig 34: Air Flow System

# Chapter 5

# AUTOMATIC/MANUAL OPERATION

## 5.0 Reciprocating Compressor Control

Reciprocating air compressors for main start air duties may be controlled manually or automatically. Automatic control is becoming popular and practical on new buildings with unmanned engine room notation which is selected for the majority of sea going vessels. On smaller coastal and offshore supply vessels unmanned engine room notation is also often requested.

### *5.0.1 Methods of Unloading Compressors*

There are several proven methods used to unload the reciprocating air compressor cylinders, and the following list gives some methods used:

a) throttling of the compressor suction.
b) compressor speed control.
c) depressors to hold the suction valve plates on their seats.
d) bypass with discharge to suction.
e) step unloading of the cylinders in multi-cylinder machines.

As marine starting air compressors are relatively small in comparison to some industrial and offshore units, methods c) and d) are usually used, and the compressor will operate in an 'on' or 'off' load condition. Depression of suction valve plates takes less power consumption when running unloaded than the system of bypassing back to the suction and atmosphere via the suction filter/silencer. However type e) can be manufactured in a modular form, it is more robust and is easier to understand.

The automation of marine air compressors is considered firstly in remote mode and secondly in automatic operation. This begins with a review of all the functions involved in local manual operation and supervision.

### *5.0.2 Starting Procedure*

Before starting an air compressor the drain valves must be opened to give an unloaded start to allow the machine to purge itself of any accumulated moisture as soon as the machine starts to turn. After a short period of running unloaded (approximately 15-20 seconds), the drains are closed and the machine is brought onto load.

### *5.0.3 Normal Running*

During compressor operation it should be periodically drained of oil and moisture that may have collected in each stage approximately every 15 to 20 minutes dependent on engine room ambient conditions. The machine should be checked to ensure that it is receiving adequate cooling water or correct air flow across the machine. If the machine has forced oil lubrication the oil pressure should be within the manufacturer's parameters.

Marine compressors are supplied with the necessary equipment to carry out the above mentioned functions. If these functions are carried out automatically operation and protection is superior to units started/stopped and observed manually. Manual operation gives reliance to a human operator with other commitments who may fail to open stage drains at the correct time and will not continuously observe machine functions. These failings may result in reduced performance or increased maintenance.

### *5.0.4 Stopping the Machine*

Upon a machine stop signal the drains should open to purge the inter and after coolers of accumulated moisture collected during the subsequent shut down period. Simultaneously the cooling flow should be shut off manually or automatically to avoid unnecessary circulation of the cool water and the possibility of condensation leading to contamination of the oil.

For an air cooled unit the continuation of passing cool air across the machine has little effect on the amount of condensation.

### *5.0.5 Mechanisation of Human Factors*

Fig 35 shows a typical electromagnetic solenoid drain valve fitted to each compressor stage. When the valve is opened it will drain any collected condensation. This type of valve is also fitted after oil and moisture separators to facilitate drainage.

It should be noted that when the compressor stops the machine will automatically unload and the solenoid valve will remain in the open position ready for the next starting sequence of the compressor.

### *5.0.6 Safety Protection Equipment*

Reciprocating air compressors may be under protected. Failures in unsupervised manual mode are more serious than when supervised or under automatic control. To meet survey requirements the monitoring of the following areas are stipulated.

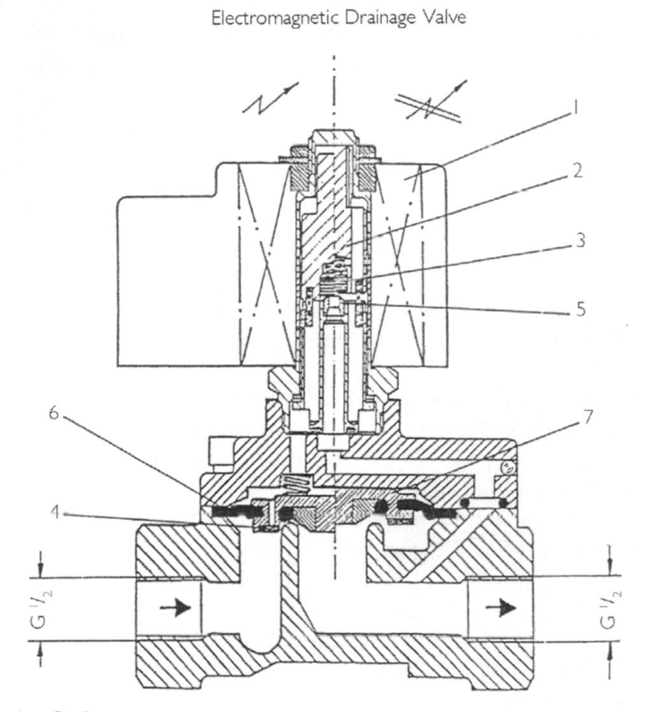


1 - Coil
2 - Core
3 - Spring damped valve disc
4 - Filter for pre-controlled channel
5 - Drainage of the pre-controlled volume via a nozzle
6 - Membrane separates flow channels from space.
7 - Membrane holder with integrated 'O' ring

Fig 35: Typical Drainage Valve

a) Lubricating oil pressure (or oil level if the compressor is a splash lubricated machine).

b) The final stage delivered air temperature is monitored directly after the compressor final air delivery and it will:

*i) On water failure*
The temperature within the compressor stages will rise and on high air temperature will shut down the compressor automatically. However as the after-

cooler is very efficient it may take an hour before the cut out temperature on the sensor is reached. In other words the siting of the high temperature cut out switch after the compressor discharge will protect the air system and not the compressor.

If the operator needs compressor protection a high temperature switch should be installed between the two final stages.

*ii) High temperature due to valve failure*

If the air temperature signal probe is fitted directly after the air compressor it will monitor the maximum temperature point after the compressor discharge and therefore protects the system itself. High temperature within the compressor will not be monitored and failure may take place before the machine cuts out.

*iii) High air temperature cut out switch before final compression stage*

Upon water failure or an unexpected rise in temperature within the machine (perhaps due to valve failure), the switch will cut out the compressor in a shorter time and so protect the compressor rather than the system.

Fig 36 shows a typical high air temperature cut off switch which should be manufactured to protection class IP65 with a service current of approximately 6A.

*iv) Air cooling failure*

In the case of fan failure in an air cooled machine a temperature probe fitted after the aftercooler will suffer a delay in responding to temperature increases within the machine. This is especially true in temperate climates as the cooler has a tropical rating and is designed to operate under those conditions. To protect the compressor the air temperature probe should be fitted between the final two stages of compression.

*v) Unbalanced load between stages resulting in high air temperature*

This can be caused by any of the following:

- Broken interstage valves.
- Carbon blocking the valve openings.
- Clogging of the air inlet filter or interstage cooler.
- Broken or partially blocked valves will cause air back flow so that the air discharge will be diluted with hot air resulting in a higher delivered temperature.
- Partial blockage of the air inlet filter (with paint or dirt), of the interstage cooler (machine running too hot – insufficient cooling water/air or oil filter blocked/heavy painted air cooler), will cause a higher compression ratio due to throttling causing reduced inlet pressure to the cylinder.

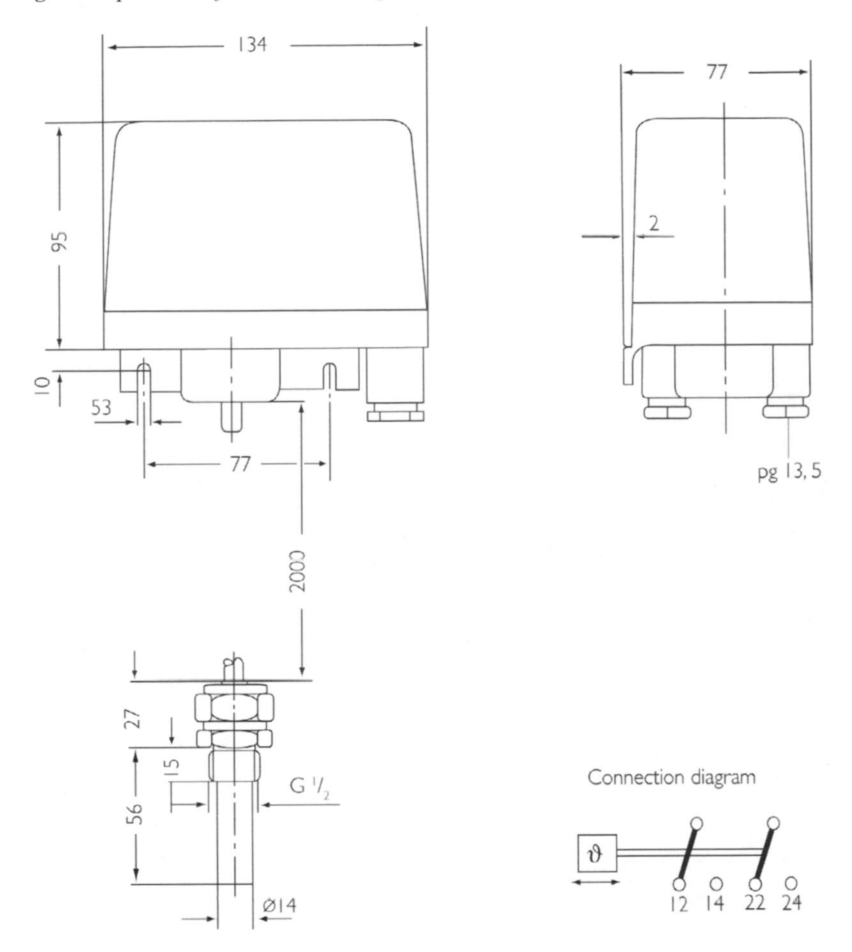


Fig 36: Temperature cut off switch

## 5.0.7 *Automatic Operation*

The preferred automatic operation of the main starting air system is for the lead starting air compressor to run in continuous mode during a vessel's manoeuvring condition with automatic loading and unloading according to the air demand. If the lead unit cannot cope with the system demand the lag (or back-up) compressor will start automatically. Under falling air pressure in the 'at sea' condition the lead compressor would again start and stop automatically according to air demand. It is essential to have a changeover switch fitted in the control panel so that the lag compressor may be selected as the lead compressor to ensure that both compressor units run approximately the same number of hours during a twelve month period. For determination of individual compressor running hours a running hours meter should be installed in the control panel – this is an important aid to maintenance schedules (see Chapter 9). Fig 37 shows a typical reciprocating compressor control panel layout.

Fig 37: Typical Control Panel

Controls etc in the compressor panel will include generally the following:

| Selector switch | Compressor 1 - 2 |
|---|---|
| | Compressor 2 - 1 |
| Selector switch | Local - Remote |
| Selector switch | Manual - off - automatic |
| Indicator lamp | Temperature |
| Indicator lamp | Oil pressure |
| Indicator lamp | Over-current |
| Indicator lamp | Operation |
| Indicator lamp | Motor heater - if fitted |
| Running hours counter | Each compressor |
| Ammeter | |

The standby machine operation will be identical to the lead machine except that the pressure switch settings will be lower.

- Compressor starts to fill the air receiver from atmospheric pressure to 30 bar.
- Compressor stops.
- On falling pressure in the air receiver the compressor will restart at approximately 27 bar.
- If the air pressure in the air receiver continues to fall then at approximately 24 bar the standby machine will start and supplement the pressure until 30 bar is reached when both compressors will stop.
- The sequence of the original operation will then continue.

The lag machine will only become operational on failure of the lead machine or if there is an exceptional heavy air demand. This system can also be adapted for multiple compressor installations. The operation of the compressor system would be as follows:

### *5.0.8 Stop/Start*

a) The compressor is stationary and in the unloaded condition – no air pressure should be within the machine.

b) The compressor is started and runs without pressure within the cylinders until the motor reaches synchronous speed and the machine will come on load.

c) If a compressor runs for long periods ie, filling the air receivers from zero to maximum operating pressure the machine will automatically unload approximately every 15 to 20 minutes depending upon the solenoid valve time settings and the prevailing ambient conditions.

d) When the compressor stops it will automatically unload and be ready for the next duty cycle.

### *5.0.9 Running Unloaded*

a) When the compressor is running on load the pressure within the machine will gradually build up until the desired pressure is reached.

b) Once the required air receiver pressure is reached the pressure switch will energise the solenoid valve. The air supply will then be cut off which will allow the trapped air to bleed off to atmosphere and enable the machine to run unloaded. A reciprocating compressor should not run unloaded for long periods, ie, a maximum of 10 minutes.

c) As air is used from air receivers for various downstream consumers (or due to system leakage), the pressure will drop in the receiver until the cut in pressure of the compressor is reached.

If air dryers are installed between the compressor and the receiver they will use between 10 and 15% of the air flow for purge air and it is important that this is taken into consideration for sizing both compressors and dryers.

d) The pressure switch will de-energise the solenoid valve, the unloader will cease to function and the machine will come back onto load and start to refill the receivers.

### *5.0.10 Notes*

i) For stop/start operation the pressure switch is connected directly to the compressor starter and the solenoid valve is then de-energised and the air line is open from the first stage delivery to the unloader.

ii) For running the machine unloaded the pressure switch is connected directly to the solenoid valve.

iii) With pressure switch settings no time delay is necessary for lagging machines due to the pressure difference. If the pressure settings are reduced a timing relay is required.

The compressor's manual or automatic equipment as described above is usually fitted by the air compressor manufacturer except for the air pressure switches which are supplied loose and installed as close to the air receivers as possible. Equipment installed within the air compressor control panels, ie, starters, selector switches (lead/lag machine and running stop/start operation), running and warning lights, running hour meters, etc, may be supplied by the compressor manufacturer or the shipyard.

Although a control panel may be installed locally at the compressors, on vessels with unmanned engine room notations it is probable that the compressor starters are installed in the group starter panel. Duplication of some controls such as running and warning lights may be fitted in the main engine room control console.

When planning an air compressor installation with fully automatic operation it is important to consider the electrical running loads that will be generated under manoeuvring conditions when the compressors may be started frequently and hence the units must be sized accordingly. Under 'at sea' conditions the compressor units will normally only be required to top up the air receivers and starting up a large compressor (say of over 30 kW), the generating budget may not take too kindly to the random starting of such a large motor.

Instead of operating a large compressor in the 'at sea' condition it can be argued that a smaller, separate compressor unit may be installed for topping up the receivers. Experience has shown that the sizes of such compressors are constantly under estimated and they must be given careful consideration.

## 5.1 Screw Air Compressor Control

Today's screw air compressor package be it a lubricated, oil free unit, air or water cooled, will likely be controlled by a microprocessor based control system to give early warning of any potential failure as well as comprehensive fault diagnosis.

The majority of screw compressor units are fitted with a silencing canopy and the control panel is situated within the enclosure. The panel should be designed so that observing and operating the controls is as simple as possible by the use of an easy-to-read plain text display. Fig 38 shows a typical display control panel layout.

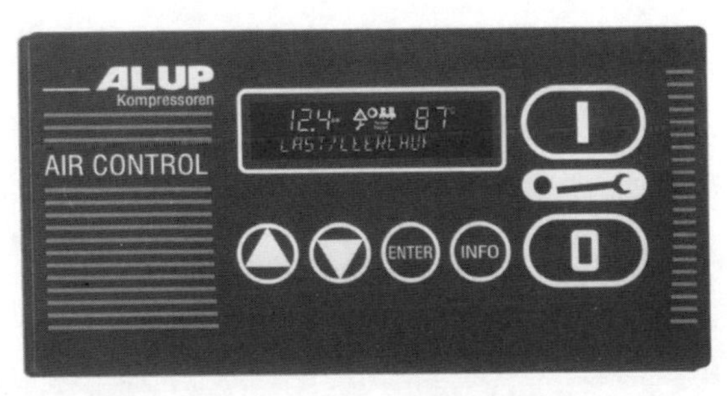


Fig 38: Screw Compressor Unit Control Panel

A typical display panel would include the following:

| Network pressure |
|---|
| Compressor discharge temperature |
| Operating mode - operating status |
| Cut-in/Cut-out pressure |
| Safety temperature settings |
| Maximum compressor discharge temperature |
| Total hours run |
| Total hours under load |
| Remaining life of air filter, oil, oil filter, oil separator cartridge: early warnings for replacement. |

Service functions should include:
- Early maintenance and fault diagnosis with fault location display.
- Automatic restart option after power failure.
- Memory for restart to initial settings after operation error.
- Protection against incorrectly entered data and commands.

Automatic shutdown if:
- Network pressure is too high.
- System pressure is too high.
- Motor temperature is too high
- Compression temperature is too high.
- Ambient temperature is too low.
- Oil pressure is too low.

If required a remote repeater control panel may be fitted in the engine control room. Alternatively if the compressor is shut down because of a malfunction then this may be repeated at the main machinery console with a remote voltmeter, signal light and audible alarm.

# Chapter 6

# SYSTEM FITTINGS

## 6.0 General

Air treatment is necessary to provide the compressed air system with a desired quality at the point of storage or use. A good air quality will help to avoid damage to pneumatic equipment. Compressed air can be contaminated by water vapour, condensate, particulate matter (either airborne or pipe scale), oil in vapour or liquid state and microbes.

Ambient air contains about 12.5 grams of water for every $1m^3$ of saturated free air at 15°C. Provided the free air is maintained at a temperature of about 15°C the water will remain as a vapour. If the air is allowed to cool below 15°C or the air is compressed water will be condensed. The temperature at which the water condenses is known as the dewpoint.

The rise in air temperature within the compressor will generally prevent any condensation but when this air passes through the aftercooler any water in the air will condense. The temperature of the air after the aftercooler should be about 20°C above the ambient and water content will need to be reduced. Water or water vapour will remain in the air and if the air temperature falls further condensation will take place.

It goes without saying that if the ambient temperature in the engine room is above 15°C the final delivered air temperature will rise more or less pro-rata.

To significantly remove more water from the delivered air than that achieved by aftercooling alone it is necessary to install an air dryer in the air line after the compressors but before the air receivers, or any other downstream equipment.

## 6.1 Air Dryers

To achieve the best quality of air it must be as clean as possible, have a low oil carry over (ie, of less than 1ppm), and undergo maximum moisture removal after each compression stage. The fitting of an oil and moisture separator after each of the final two stages of the compressor will reduce the oil and moisture carryover into the air system. If drier air is required an air dryer should be fitted after the compressor. Several different types of air dryer are manufactured with the most common types being described in the following sections.

### *6.1.1 Desiccant Absorption Dryer*

To achieve clean dry air at the point of use, heated or heatless type twin tower, desiccant air dryers with a special chemical desiccant and utilising drying cycles may be employed. The dryers would be installed in parallel and have a prefilter. A final dust collecting filter would be installed directly after the dryer itself. Both filters would normally be included and supplied within the dryer package. Fig 39 shows a typical absorption dryer layout.

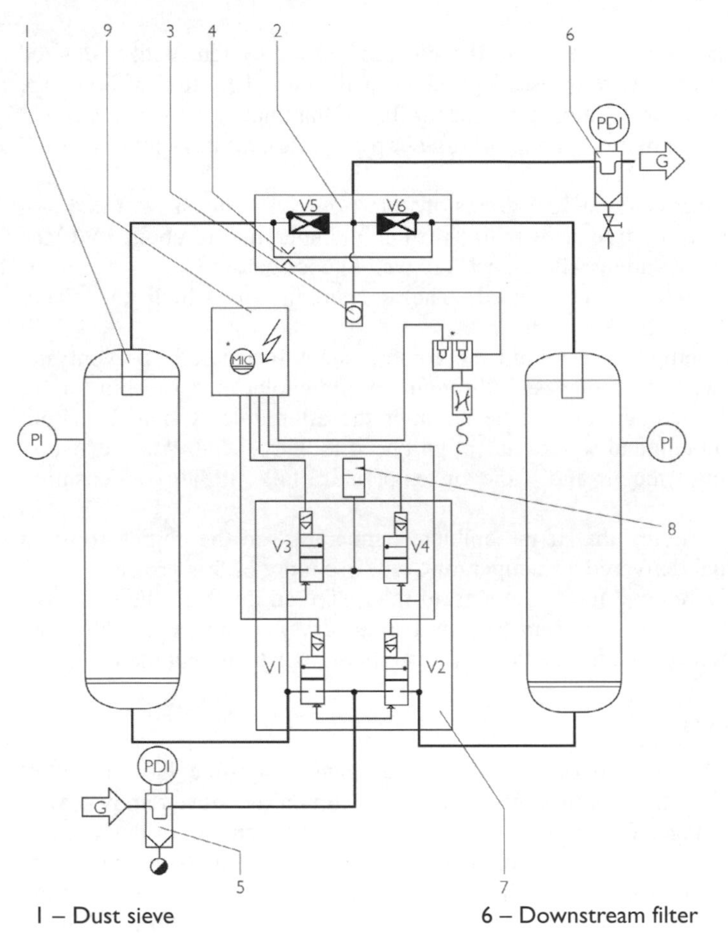


1 – Dust sieve
2 – Check valve block (V5-V6)
3 – Humidity indicator
4 – Regeneration gas nozzle
5 – Upstream filter
6 – Downstream filter
7 – Solenoid valve block (V1-V4)
8 – Silencer
9 – Control cabinet
*Optional: Ecotronic control system with dew point measuring unit

Fig 39: Absorption Dryer Schematic Diagram

### *6.1.2 Loading*

The air to be dried enters the operational dryer B1 via the pre-filter and the 4-port directional valve combination V1 (V2). By means of the lower wedge wire sieve, the air is distributed within the desiccant and the air flow is upwards. The desiccant will absorb the water vapour contained in the air. The air will leave the desiccant bed when the desired dew point has been reached. At the outlet the dried air will flow into the upper control block with the integrated non-return valve V5 (V6) to the activated carbon absorbent. Fig 40 shows a process diagram of the system.

### *6.1.3 Regeneration*

When the loading has been completed the dried air control will switch over and the loaded desiccant vessel B1 will be regenerated. During the switchover the blow off valve V3 (V4) will close and after approximately 3 seconds after the switchover this valve will open. A small amount of the dried air is retrieved within the check valve block and used for regeneration. This air will pass through a bypass orifice and flow without pressure through the vessel from the top to bottom.

The quantity of air used for regeneration depends upon the operating over pressure and the temperature of the air to be dried. When sizing the drier unit it is necessary to allow for the purge air which is from 10% to 15% of the total air capacity.

The pressure of the regeneration air is lowered to the value of atmospheric pressure and it absorbs the humidity contained in the desiccant. Finally the air loaded with humidity flows out into the atmosphere after passing through the lower 4-port exhaust valve combination V3 (V4) and a silencer. The loading time will be extended if electronic dew point control is used.

By using silica gel ($SiO_2$) which is one of the most common of the absorption materials an atmospheric dewpoint of -50°C is possible with a maximum inlet temperature of 50°C.

### *6.1.4 Pressure Build-up with Switchgear*

The period of time between the end of regeneration and the change-over required for pressure build up is approximately one minute. This will enable the changeover to the other vessel B (A) to happen without any pressure shock. For this the opened blow-off valve is closed and in the regenerated vessel the pressure increases to the normal operating pressure. The pressure phase will be finished when both pressure gauges show the same pressure. The changeover from loading to regeneration is time dependent and normally approximately 5 minutes.

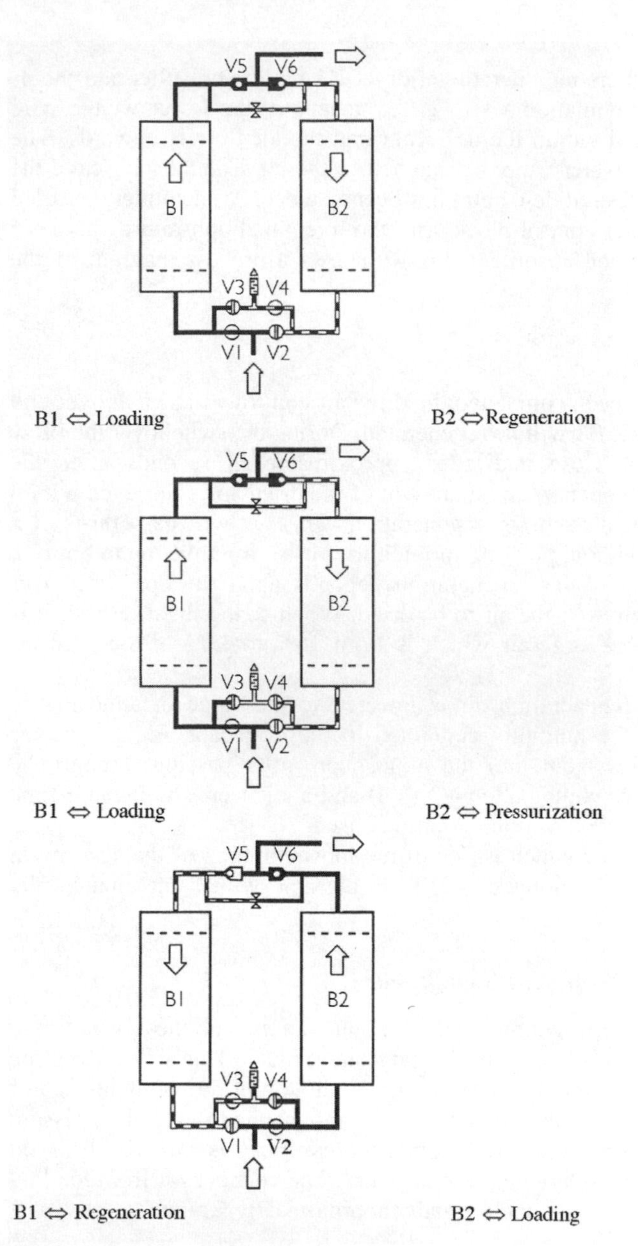


Fig 40: Dryer Process System

### *6.1.5 Dryer Alternatives*

Two alternatives that may be considered are either cold or hot regeneration. Cold regeneration has a drying and regeneration time of some 5 minutes and for this reason the moisture only deposits on the outer surfaces of the drying agent. With hot regeneration the drying and regeneration times will be six to eight hours and during this long drying period the moisture deposits on both inner and outer surfaces of the drying agent.

### *6.1.6 Refrigerant Dryers*

Complete air treatment packages of refrigeration dryers with microfilters can be supplied and sized to meet the air system requirements. The dried, filtered, compressed air may have a dewpoint of around +3°C and at a pressure of up to 13 bar (corresponding to an atmospheric dewpoint of -21°C) is clean, moisture free and ready for use. Fig 41 shows a typical refrigerant dryer.

Fig 41: Refrigerant Air Dryer

After start-up the pressure dew-point will automatically adjust itself independently of air demand variations. The dangers of freezing at low air flows will be eliminated since the temperature within the dryer is held constant.

When the compressor goes off-load the refrigeration dryer will switch off automatically which leads to additional energy savings.

The refrigerant air dryer can be installed either before or after the air receiver. The advantage of installing a refrigeration dryer over a desiccant dryer is that no desiccant materials have to be used saving maintenance time.

### *6.1.7 Dehydrator (Membrane) Dryer*

The heart of this type of dryer is thousands of polymer hollow fibres that allow some gases to permeate across the fibres whilst other gases continue to pass through the 'hollow' core of the fibre. These fibres provide maximum separation efficiency by using the difference between the partial pressure of the gas on the inside of the hollow fibre to that on the outside.

The dryer works under varying atmospheric conditions and will give a dewpoint of 40°C below the inlet temperature to the dehydrator. An advantage of this type of dryer is the savings made in not supplying and storing onboard chemicals that have to be used in the regeneration or absorption types.

The principle of operation allows 'fast' permeation gases such as water vapour to escape through the wall of the fibre leaving the dry compressed air to continue through the centre of the hollow fibre. A small amount of approximately 7% of the 'slower' compressed air (a mixture of oxygen and nitrogen) also permeates through the fibre wall to 'sweep' away the water.

The self-generating sweep or purge air eliminates the need for additional purge or control valves, regulators and flow meters. The air consumption is much less than that required by an absorption dryer and it is possible to fit a smaller unit. The negative side of using this type of dryer is that oil free air is required to avoid membrane damage. Therefore it is necessary to install a selection of oil removal filters in the line before the air enters the membrane unless an oil free air compressor is installed.

## 6.2 Air Receivers

### *6.2.1 Starting Air*

To meet the regulations of the survey classification rules a minimum of at least two starting air receivers are required. The sizing of the receivers for main starting air use will depend if the main propulsion engines are of the reversing type and are normally sized for a minimum of twelve starts. If the propulsion engine is a non-reversing engine the receivers will be sized for a minimum of six starts. The sizing of each air receiver will be calculated by using the following formula:

$$\text{Receiver capacity } (m^3) = \frac{\text{Vs x No of Engines x No of Starts (S)}}{P_2 - P_1}$$

Vs = Air required for one start ($m^3$)
$P_2$ = Maximum pressure of air receiver (bar)
$P_1$ = Minimum pressure of air receiver (bar)
S = 12 starts for a reversing engine or 6 starts for a non-reversing engine.

On a merchant ship the air receivers should be sized so that a continuous running air compressor does not come 'off load' for periods of less than five minutes. For compressors operating on a stop/start principle the duty unit should not have more than six starts in one hour. More than six starts in an hour may result in burning out the starter or motor.

Air receivers have to be fitted with an array of valves and fittings:

- inlet air valve
- outlet air valves
- drain valve
- safety valves
- handhole or manhole
- pressure gauges.

An alternative would be to fit a valve head onto the air receiver. The air receivers may be installed horizontally or vertically. If installed horizontally they should be arranged with a slight slope toward the drain cock to ensure any condensate in the receiver can be manually drained away. A preferred alternative is an automatically operated drain valve.

Air receivers must be fitted with a safety valve which is designed to handle the full flow of air with no more than a 10% rise in the air pressure. It is also advisable to have a fusible plug fitted and of sufficient size to carry out a similar function should a fire occur in the receiver which may create a malfunction in the safety valve.

Receivers also receive a certain amount of oil and moisture from the air system carried downstream from the air compressor itself. It is therefore essential that the inside of a receiver is examined at least once every twelve months and the internals recoated if necessary. Large air receivers are fitted with a manhole door so that they can be easily inspected and recoating carried out. Smaller air receivers are fitted with a small handhole.

Air receivers should be installed in a cool area wherever possible since more condensate will form inside the receiver which should be drained away from the receiver periodically otherwise the condensate will build up and ultimately pass downstream into the air distribution system. The majority of air receivers are manufactured in steel.

### *6.2.2 Working Air*

The working air (or service air) receiver capacity will depend upon the maximum number of air tools and other air consuming equipment that the vessel is expected to use at any operating time. The receiver air storage pressure will be between 7 and 13 bar depending on the ship design operating parameters.

The type of fittings on the receiver will be similar to those fitted to the main start air receiver – see 6.20 above and as the receiver is a pressure vessel it should be certified by the selected survey authority.

### *6.2.3 Control Air*

These air receivers are normally smaller than working air receivers in capacity and are designed to operate at a working pressure of between 7 and 13 bar. The receivers are fitted with similar fittings to the main air receivers. Once again the receivers should be certified by the selected survey authority.

### *6.2.4 General*

Typical air receivers are laid out as shown in Figs 42 and 43. If the receivers are sited horizontally they should be installed with a slight slope toward the drain valve.

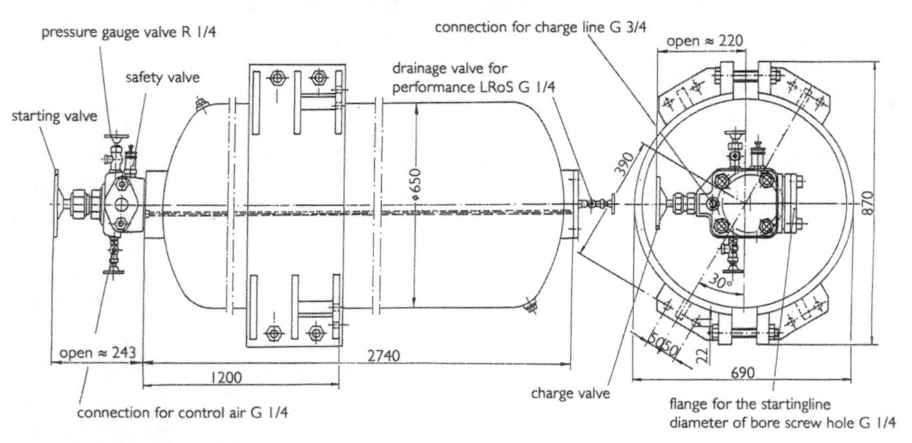


Fig 42: Air Receiver with Valve Head

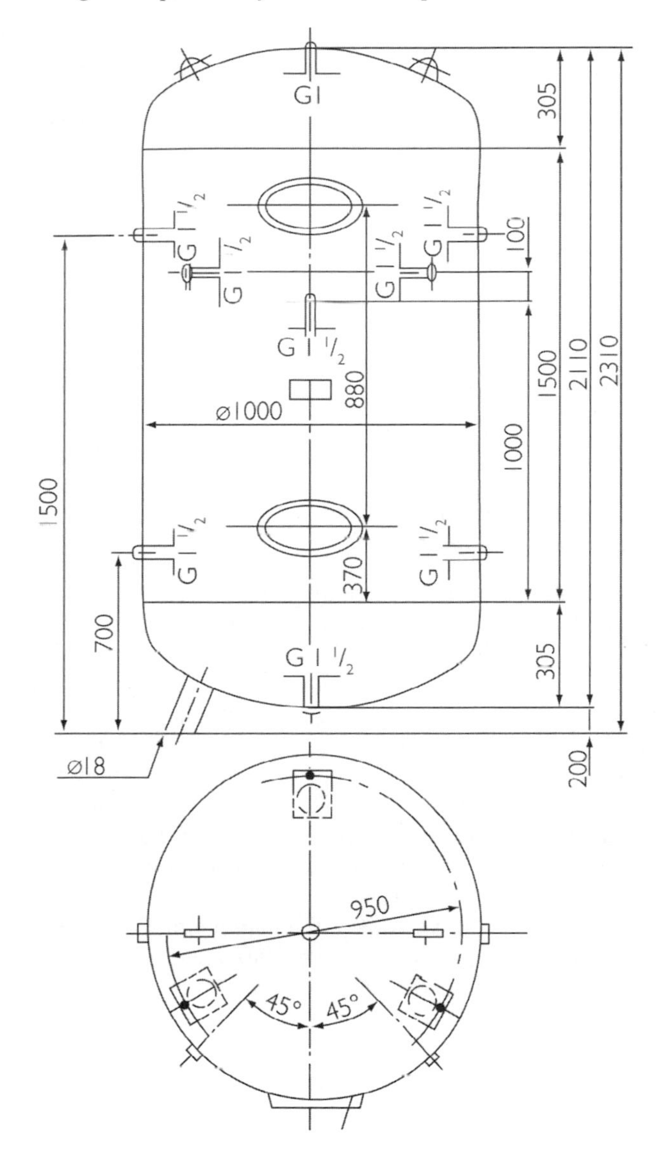


**Compressed Air Receiver**

standing design with foot
working pressure: 30 bar
content: 1200 litre

equipped with following

1 safety valve
1 pressure gauge Ø 100
1 test flange $^{1}/_{2}$"
1 drain valve G $^{1}/_{2}$
1 outlet valve G 1 $^{1}/_{2}$
1 filling valve G 1 $^{1}/_{2}$

Labels: language acc. to

Test: acc. to order

Coating: galvanized

2 screw plugs G 1 $^{1}/_{2}$

2 screw plugs G $^{1}/_{2}$

Completely delivered with all fittings and gaskets

Fig 43: Air Receiver with Separate Fittings

## 6.3 Pressure Maintaining Valve

For air compressors designed to operate at pressures up to 40 bar it is not considered necessary to install a pressure maintaining valve in the air system after the main start air compressor. The fitting of such a valve first came into prominence in compressors that had several stages of compression and operating at pressures of over 60 bar. Fitting a pressure maintaining valve after a high pressure compressor enables a compressor to rapidly reach its working pressure within the machine rather than fill up the air receiver and its air pipework stage by stage. If the valve was not fitted the compressor would have a lot of 'valve flap' which could lead to excessive wear of valve parts and piston rings resulting in increased spares consumption and higher maintenance.

If an absorption or equivalent type of air dryer is fitted it is necessary to fit a pressure maintaining valve after the dryer to ensure that the operating pressure is maintained within the unit. This valve is often incorporated in the dryer package.

## 6.4 Pipeline Non-return Valve

This valve can either be supplied as a loose item and fitted by the installer or fitted by the compressor maker directly after the compressor air discharge outlet. The valve will operate as follows:-

When the compressor unloads there will only be compressed air in the compressor body and the inter and after cooler tubes together with the short length of piping which unloads the compressor to atmosphere. This is not only economical but the non-return valve also prevents the downstream side of the compressed air re-entering the compressor and bringing with it moisture droplets which may settle on the discharge valves creating a malfunction. An inrush of air would also occur if the non-return valve was placed at a long distance from the compressor discharge.

It is recommended that the non-return valve is sited at a maximum of one metre distance from the air discharge of the compressor but after a flexible pipe if fitted.

The fitting of a non-return valve will also act as a safety feature if any maintenance work is commenced by accident without first closing down the isolation valve.

If air compressors are fitted with automatic unloaders it is essential that automatic cooler drains are fitted otherwise the machine will not be truly automatic and periodic manual draining of the compressor coolers is necessary.

## 6.5 Isolation Valve

From time to time it is necessary to carry out maintenance work on the air compressors and it is very important that in addition to the non-return valve an isolation valve is installed between the non-return valve and the air receivers. The fitting of the isolation valve will also act as a safety precaution to protect the engineer and this valve MUST be closed before any work is carried out on the compressor.

## 6.6 Oil Filters

Oil filters are only required to be installed directly after the air compressor and in the delivered air line if almost oil free air is required and for sensitive equipment eg, the control air system which would include control air receivers used for storage before the air is delivered downstream.

Always present in the atmosphere dust particles will be drawn into the compressor and increased eightfold by the compressor itself which will have to be filtered out if necessary depending upon the downstream equipment. Oil and dust tend to mutually attract each other. Fig 44 shows a section of an oil removal filter.

Fig 44: Oil Removal Filter

# Chapter 7

# OPERATING INSTRUCTIONS

## 7.0 Reciprocating Compressors

The operating and maintenance manual of the selected reciprocating compressor type will include the operating procedures. If the manufacturer's manual is not available the following will outline the general procedures to be followed.

## 7.1 Initial Start

This procedure should be carried out after the initial installation of the compressor or after any major overhaul and when accepting replacement units. Make sure that the unit is isolated from the electrical supply and that the cooling water (if this is the cooling medium), is shut off and the delivered air isolating valve has been closed to ensure that no high pressure air may find its way into the compressor whilst it is being worked on.

## 7.2 Lubrication

The compressor will be delivered to site with no oil in the sump. Only branded oil should be supplied and used according to the compressor manufacturer's recommendations. The machine sump is to be filled to the level indicated on the compressor sight glass or dipstick. A list of recommended oils will be listed in the compressor operating manual.

Oil levels should be checked daily when the compressor is in operation and if the oil level falls to the lower oil indicator sump level the oil must be topped up to the higher oil level. Over filling of the oil in the sump can lead to operating problems, ie, higher than expected oil carryover into the air system. The majority of air compressors are fitted with a low oil level cut-out switch.

## 7.3 Water Cooled Unit with Water Pump

The cooling water flow to the compressor should be adjusted so that the cooling water outlet has a temperature differential of between 10°C and 12°C above the cooling water inlet temperature.

If the water pump is driven directly by the compressor when the unit stops the pump will also stop and no water flow will enter the compressor reducing the possibility of condensation formation. If the water pump has its own electric motor then this should be connected into the compressor starter so that the pump will only start and stop at the same time as the compressor.

## 7.4 Water Cooled Unit without Water Pump

The control of the water shut off valve sited at the cooling water inlet must be monitored. When the compressor is not running the valve must be closed and this can be carried out either automatically or manually. When the compressor is running the water flow should be adjusted so that the water temperature rise across the machine is 10°C to 12 °C to obtain the optimum cooling conditions for the compressor

## 7.5 Unloading

All of the cooler drains should be opened together; unloading may be carried out manually or automatically.

## 7.6 Gauge Cocks

The lubricating oil gauge cocks should be fully open when the compressor is running but the air gauge cocks should only be partially cracked open to avoid fluctuations resulting in gauge damage.

## 7.7 Air Intake

The air inlet filter should be checked to ensure that it is in a clean condition. Heavy painting on the filter insert is not unknown especially when a vessel is new or when machinery has recently been painted. If paint contamination has taken place the filter insert must be changed or damage to the compressor may ensue.

## 7.8 Relief Valves

Ensure that the valve springs will operate ‘freely’.

## 7.9 Final Compressor Check

The compressor should be turned by hand for several revolutions to ensure that the machine is rotating freely.

## 7.10 Air System

Check that the air system downstream of the compressor is correctly connected.

## 7.11 Compressor Area

The immediate area around the compressor and on the machine itself should be clean and clear of all loose particles.

## 7.12 Initial Machine Operation

Run the compressor for a few revolutions to ensure that it has the correct rotation; a label showing the correct rotational direction should be displayed on the unit. If the rotational direction is correct the machine can be restarted under manual control.

When the machine has reached its design running speed the oil and interstage air pressures should be compared with the manufacturer's recommendations listed in the operating manual. If the machine is running satisfactorily the drain and unloading valves should be closed slowly and the unit brought onto full load.

For compressors fitted with manual drains the valves should remain slightly cracked open to avoid condensation which may build up within the machine.

All pressure gauges should show the normal working pressure of each stage within a 20 to 30 second period after the drain and unloader valves have been closed.

## 7.13 Stopping the Compressor

For manually controlled compressors after a unit has stopped all drain and unloader valves should remain open for two to three minutes and the cooling water flow stopped.

If compressors with automatic drain and unloading valves run for long periods they should be fitted with automatic cooling water stop valves. The reason is to ensure that the water will not flow through the machine when it is stopped since this may cause excess condensation within the machine.

## 7.14 Running In

Modern air compressors will be tested under full load conditions before leaving the factory with the final test being witnessed by the local class surveyor and possibly by the owners or shipbuilders. Apart from the procedures mentioned above when the compressors are installed onboard no special running in periods would be anticipated unless specified by the shipbuilder or owner.

## 7.15 Not In Use

Compressors not anticipated to be in use for periods of over four weeks should be run for about 30 minutes in each four week period. Additional corrosion protection is unnecessary during a service interruption of less than twelve weeks.

For a stoppage of more than twelve weeks oil in the compressor sump must be drained and replaced by a recommended anti-corrosion oil in accordance with the manufacturer's manual. As the anti-corrosion oil has adequate running characteristics the compressors may be operated with this type of oil for short periods.

## 7.16 Corrosion Protection

The following procedures may be adopted for corrosion protection of the compressors when they are not in use:

- Operate the compressor without pressure with the drain valves open for about five minutes in order to blow off any collected moisture.
- Drain the compressor oil from the crankcase sump.
- Replace the oil in the sump with the manufacturer's recommended anti-corrosion oil to about 90% of the maximum level.
- Start the compressor without pressure with open drain valves and allow the machine to run for five minutes before stopping the machine.
- Remove the compressor air filter, restart the compressor, again without pressure and slowly inject the remaining 10% of the anti-corrosion oil through the air filter opening.
- Stop the compressor when oil mist starts to leave the air outlet.
- Refit the air filter.
- Drain the anti-corrosion oil from the crankcase sump.

## 7.17 Tightening Values

It is important that the nut/bolt tightening values are listed preferably in a tabular format in the maintenance section for each major component ie, crankshaft, cylinder covers, etc.

## 7.18 Standard Factory Acceptance Tests

The following is a typical procedure for testing the air compressor units before delivery.

### *7.18.1 Check of Documents*

All documents should be checked to ensure that the customer requirements for testing the compressor are met.

### *7.18.2 Running In Programme*

The compressor should be run on the manufacturer's test bench for a period of 16 hours (depending on the maker's standard), with incremental increases in stage pressures.

### *7.18.3 Areas to be Observed*

The following areas are to be observed during the test:
- Visual check for damages and leakages
- Leak test
- Functional test
- Performance test as per contract
- Control of safety devices eg safety valve on last compression stage
- If required issue factory acceptance certificate.

### *7.18.4 Acceptance Test*

If an acceptance test is required to be witnessed by the customer and the survey authority the following should be carried out:
- Check all of the documents relating to the test requirements.
- Carry out the performance test as per the contract
- If required carry out examinations according to special test specification.

### *7.18.5 Test Certificate*

The test certificate is signed by the manufacturer and class surveyor and will contain information such as test pressures, safety valve setting pressures, compressor test operating times, performance test data, and if specified the crankshaft test and any other special customer requirements.

### *7.18.6 Corrosion Protection*

If it is deemed necessary for the compressor to have corrosion protection then this will be carried out in accordance with 7.16 above.

## 7.19 Safety

Safety is obviously very important when operating air compressors. The details in Fig 45 give good general guidance and depict the safety markings on the machine.

## Safety markings on the machine

**Caution!**

Safety markings affixed to the machine must not be altered or removed. Replace damaged or lost safety markings immediately, true to the original.

Sauer compressor models with an EC Manufacturer's Declaration or EC Declaration of Conformity are marked with the following safety markings.

| Safety marking | Meaning |
|---|---|
| | Danger! High volatge! |
| | Compressor starts automatically without warning! |
| | Hot surface! |

| Safety marking | Meaning |
|---|---|
| | Read instructions! |
| | Wear hearing protection! |
| | Rotational direction of crankshaft |

**Location of safety markings (top view)**

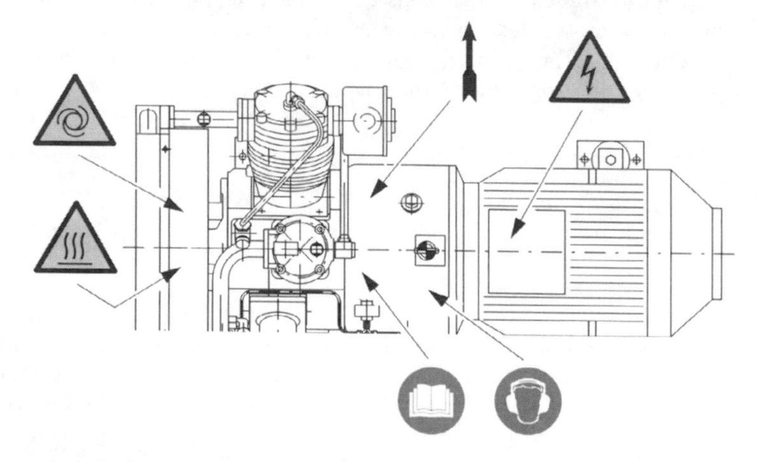

Fig 45: Safety markings on the machine

# Chapter 8

# OILS

## 8.0 General

For many years reciprocating starting air compressors used engine oil or equivalent for lubrication mainly because of shipowners' preference to keep the number of different types of oils stored onboard to a minimum. Recent years have seen the introduction of synthetic and special mineral oils for air compressor use which have now been accepted by owners. This change has increased the life of compressor components and improved reliability of the various types of machines including air and water cooled units. The only disadvantage to the shipowner is that he has to carry and dispose of an additional oil grade but this is a cost worth paying for the many advantages.

## 8.1 Synthetic Oils

Oil companies are now promoting the use of synthetic oils for use in both reciprocating and screw compressors. There are both advantages and disadvantages to using synthetic oils some of which are described below.

### *8.1.1 Advantages*

a) The high degree of thermal stability of synthetic lubricants reduces the incidence of coking up the compressor's suction and delivery valves, cooler tubes and pipework. This will keep the compressor running parts operating much longer than previously experienced. Seals made from fluorinated hydrocarbon, silicone, fluorosilicone, viton R, teflon and high nitrile Brune (N NBR) are compatible for use within the compressor eg, concentric valve seals.

b) Large capacity (60 $m^3/h$ and above) two stage units having delivery pressures in excess of 20 bar have high final air delivery temperatures due to their fundamental design. Three stage air cooled units have lower interstage temperatures that are lower than two stage water cooled compressors.

At high air temperatures of above 230°C synthetic oils have a greater chemical stability which helps reduce oil ageing. Even at temperatures as low as -30°C synthetic oils may be used although at this low temperature lower viscosity oil would have to be selected. Fig 17 (Chapter 3) illustrates typical flash points of both synthetic and mineral oils.

c) Seagoing vessels experience a wide variation of climatic conditions which can affect the running conditions of the machine. Synthetic oils do not have to be changed as often due to varying climatic temperatures.

d) Based on available toxicological information most synthetic oils produce no adverse effects on operator's health when properly handled and used. No special precautions are considered necessary beyond the attention to good personal hygiene including laundering of oil soaked clothing and washing any skin contact areas with soap and water. Additional health and safety information is available from the oil supplier.

The following table relating to Mobil grades illustrates typical synthetic oil characteristics.

| **Mobile Rarus** | **Test method** | **427** | **428** |
|---|---|---|---|
| ISO Viscosity grade | | 100 | 150 |
| Viscosity cSt at 40°C | ASTM D445 | 107 | 145 |
| Viscosity cSt at 100°C | ASTM D445 | 10.1 | 12.8 |
| Viscosity index | ASTM D2270 | 66 | 75 |
| Specific gravity | ASTM D1298 | 0.958 | 0.970 |
| Flash point, °C | ASTM D92 | 270 | 270 |
| Pour point, °C | ASTM D97 | -36 | -40 |
| Rust protection | ASTM D665 | | |
| Distilled water | | pass | pass |
| Foam test | ASTM D892 | | |
| Sequence I, ml | | 10.0 | 50.0 |
| Air release | ASTM D3427 | | |
| Minutes to 0.2% | | 4 | 2 |
| Colour | ASTM D1500 | 3.0 | 3.0 |

e) Component life is greatly improved ie, the suction and delivery valves are claimed to have a life of up to eight times above that previously obtained using standard engine oil.

f) The intervals for changing synthetic oil are almost six times longer than achieved using standard engine oil.

### *8.1.2 Disadvantages*

a) Synthetic oil is hygroscopic and has a strong urge to absorb any water that will be present in the crankcase due to condensation. This reduces the oils ability to act as a lubricant and increases the possibility of a piston seizure if not used correctly.

b) The crankcase must be clean of any paints such as lacquer, varnish, pvc and acrylic paints since synthetic oil acts as a paint stripper. Paint particles peel off passing into the compressor internal air flow system and downstream of the compressor clogging up vital parts of the unit and the air system within the compressor with bad effect. Such particles may also block the lubrication system.

c) Early production synthetic oils had low viscosities and if left in an idle compressor for several weeks had a tendency to drain away from the sides of the cylinder walls. In this condition there is a strong possibility that oxidization would take place and on restarting the compressor damage to cylinder walls and piston rings could occur. Such damage could be avoided if the compressor was turned over every four to five weeks. Due to this problem oil companies have done further research and now offer oils which cling more readily to the cylinder walls reducing the possibility of oxidization.

d) Cost can also become a major influencing factor in selection insofar as synthetic oils may cost up to six times more than that of a compressor mineral oil. If the unit has a running time of say 2000 hours per year, then a compressor with an absorbed power of 50 kW would have an additional oil cost of about US $ 500 per year to add to the maintenance bill.

e) There is a lower availability of synthetic oils world wide.

f) If a compressor using synthetic oil is idle for several weeks there is a strong tendency for oil and water separation. It is essential to drain off this free water collected at the bottom of the sump before starting to avoid the possibility of piston seizure due to lack of oil film on the cylinder walls.

Synthetic and mineral oils should not be mixed and if synthetic grades are introduced later the crankshaft and other lubricated components must be cleaned free of mineral oil. If the compressor is operated with synthetic oil from the outset the manufacturer's instruction manual should be followed. Many manufacturers state that the compressor should have an initial running in period of approximately 100 running hours using an approved mineral oil to assist in the 'bedding in' of the compressor.

## 8.2 Mineral Oils

Following the introduction of synthetic oils many oil companies introduced mineral oils as an alternative. These oils have been designed primarily for use in reciprocating compressors. An important advantage of this oil type is a lower purchase price compared to synthetic oils. This could be an important consideration in the overall running costs of the compressor. Mineral oils are easier to recycle whereas synthetic oil process and disposal may take several days even when correct equipment is available.

Component life using this type of mineral oil can be at least five times greater than previously attained using earlier mineral engine oil grades.

Compressor mineral oils are a high quality lubricant and are formulated from refined base stocks which have a narrow distillation range. They incorporate novel additive technology to give outstanding anti-wear, antirust and anti-oxidation performance.

The advantages of using a modern compressor mineral oil include:

- Cleaner compressors as a result of lower deposit formation over conventional mineral oils giving longer running periods between maintenance shutdowns.
- Less wear because of outstanding anti-wear properties and protection against corrosion.
- Extended lubricant life due to exceptional resistance to oxidation and thermal degradation and reduced carryover.
- Good inter-cooler and after-cooler efficiency due to reduced deposits and high resistance to emulsion formation.
- Effective for continuous high temperature operation up to 220 °C.

Similar comments with regard to health and safety apply as per synthetic oils detailed earlier.

The following table illustrates the characteristics of a typical modern compressor mineral oil.

| **Mobil Rarus 426** | **Test method** | **426** |
|---|---|---|
| ISO VG Viscosity | ASTM D445 | 68 |
| Viscosity cSt at 40°C | | 65 |
| Viscosity cSt at 100°C | | 8.7 |
| Viscosity index | ASTM D2270 | 95 |
| Specific gravity | ASTM D1298 | 0.875 |
| Flash point, °C | ASTM D92 | 252 |
| Rust protection | ASTM D665 | |
| Foam test | ASTM D892 | |
| Foam sequence 1, ml | | 50/0 |
| Colour | ASTM D1500 | 1.5 |

## 8.3 Oil Separation

Air compressors require reliable types of oil lubrication to ensure that the machine is reliable and has a high service life. The compressed air condensate will contain finely dispersed and emulsified oil. Light liquid separators or oil settling tanks after a moderate settling time of several hours will still contain water with a residual oil content of more than 30 mg/litre. The maximum tolerable load is 20 mg/litre and in some instances it may be even lower than this value.

If values of below 5 mg/litre are required a fully automatic oil separation system will be required. The oily condensate will flow into the chamber with or without a low pressure. Moderation will provide for the floating of undissolved and non-emulsified oil to the surface of the separator or tank where it will discharge via an adjustable overflow pipe to a collecting tank. The purified water passes through a specially activated carbon filter to absorb any emulsified oil residues and finely dispersed oil droplets.

## 8.4 Oil for Screw Compressors

Synthetic oils for screw compressors have been the standard almost from when this type of compressor was introduced and accepted by shipowners in the 1980s.

The rotors of a screw compressor will have a speed of up to 10 000 rev/min and operate at pressures of up to 13 bar without the need to worry about its reliability. Figs 46a & b shows a screw compressor flow diagram including the cooling fluid (oil).

## 8.5 Oil Content in the Air

See Chapter 3 section 3.24.

## 8.6 Own Brand Oil

Several compressor manufacturers have decided to introduce an ‘own’ brand oil following extensive testing with a selected oil company. Whereas reliability may have advantages guarantees may not be honoured if this type of oil is proved not to be have been used with the compressor in service. Such oil may prove difficult to obtain worldwide.

## 8.7 Oils General

A list of manufacturer’s recommended oils are included in the compressor operation manual. Tables 8.1a & b show lists of typical lubricant suppliers and grades.

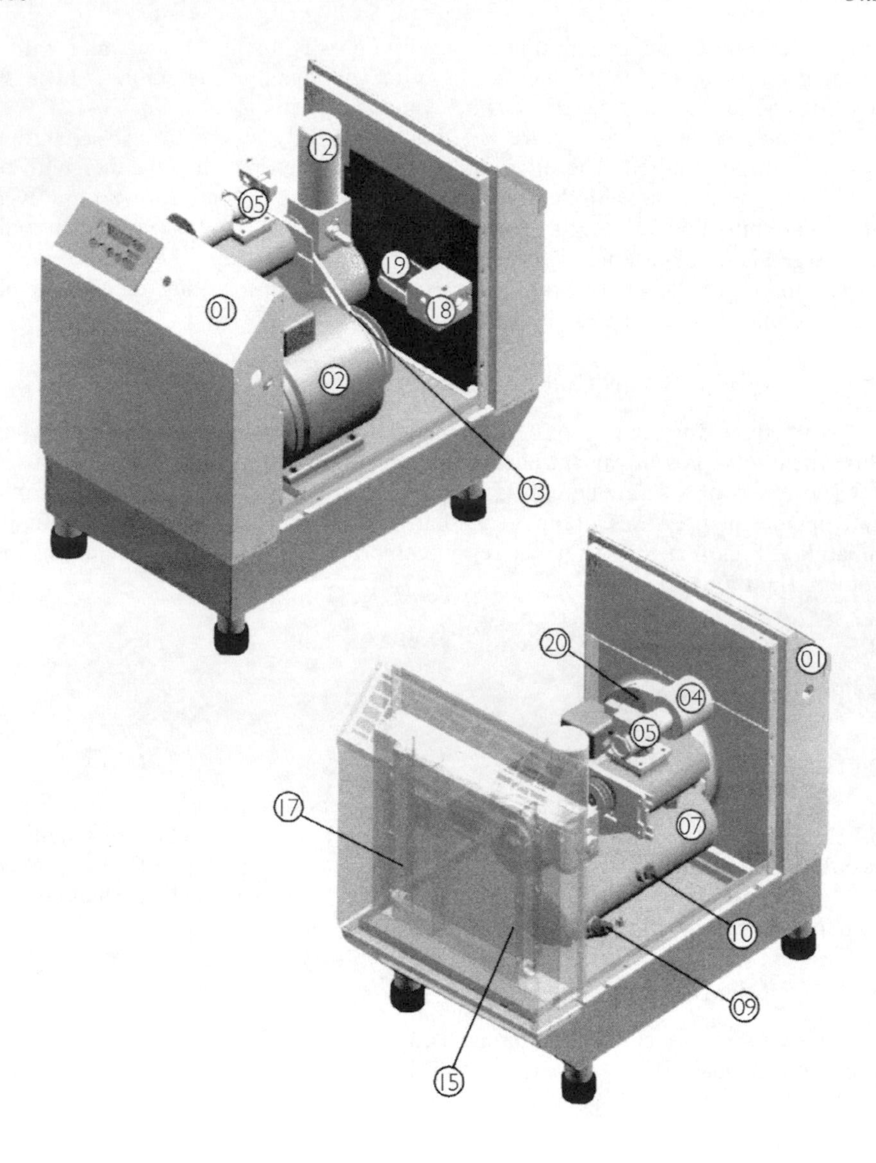


Fig 46a: Screw Compressor Flow Diagram

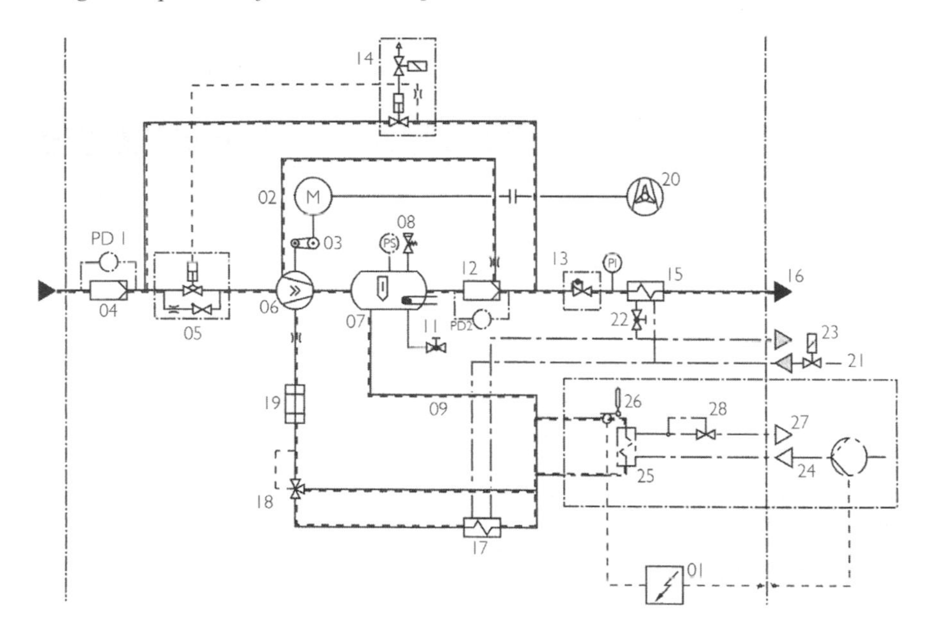


1 – Switching cabinet
2 – Electric motor
3 – V-belt
4 – Air filter
5 – Suction regulator
6 – Compressor level
7 – Cooling fluid reservoir
8 – Safety valve
9 – Drain
10 – Inlet fitting
11 – Additional heater (optional)
12 – Cooling fluid separator
13 – Minimum pressure non-return valve
14 – Relief valve
15 – Compressed air after-cooler
16 – Compressed air connection
17 – Cooling fluid cooler
18 – Cooling fluid temperature regulator
19 – Cooling fluid filter
20 – Cooling air ventilator
21 – Cooling water inlet
22 – Throttle valve compr air after-cooler
23 – Solenoid valve
24 – Water inlet
25 – Heat exchanger
26 – Thermostat
27 – Water outlet
28 – Thermostat valve

Fig 46b: Screw Compressor Flow Diagram

Table 8.1a: List of typical lubricant suppliers and grades (Lubricating Oils)

| Brand | Name | Group |
|---|---|---|
| Agip | Diesel Gamma 30 | VCL-100 |
| | Dicrea 100 | VDL-100 |
| | Acer 100 | VCL-100 |
| | Motor Oil HD 30 | SAE 30 |
| | Cladium 50 | SAE 30 |
| ARAL | Motanol HE 100 | VDL-100 |
| | Kowal M 30 | VCL-100 |
| | Disola M 30 | SAE 30 |
| AVIA | Avilub Compressor Oil VDL-100 | VDL-100 |
| | Avilub Compressor Oil VDL-100 | VCL-100 |
| | Motor oil HDC 30 | SAE 30 |
| | Motor oil HD 30 | SAE 30 |
| BP | Energol RC 100 | VDL-100 |
| | Energol IC-DG 30 | VCL-100 |
| | Energol DL-MP 30 | SAE 30 |
| | Energol OE-HT 30 | SAE 30 |
| | Vanellus C3 SAE 30 | SAE 30 |
| Castrol | Aircol PD 100 | VDL-100 |
| | Marine CDX 30 | SAE 30 |
| Chevron | HD Compressor Oil 100 | VDL-100 |
| | Delo 1000 Marine 30 | SAE 30 |
| | Veritas 800 Marine 30 | SAE 30 |
| | RPM Heavy Duty Motor 30 | SAE 30 |
| DEA | Actro EP VDL-100 | VDL-100 |
| | Trion EP VDL-100 | VDL-100 |
| | Regis SAE 30 | SAE 30 |
| Esso | Exxcolub 77 | VDL |
| | Exxcolub 100 | VDL-100 |
| | Compressor Oil 3021 N | VDL-100 |
| | Exxmar 12 TP 30 | SAE 30 |
| | Exxmar XA | SAE 30 |
| | Essolube HDX Plus +30 | SAE 30 |
| Mobil | Rarus 427 | VDL-100 |
| | DTE Oil Heavy | VDL |
| | Mobilgard 300 | SAE 30 |
| Shell | Corena Oil P 100 | VDL-100 |
| | Rimula X 30 | SAE 30 |
| | Melina S Oil 30 | SAE 30 |
| | Melina Oil 30 | SAE 30 |
| | Gadinia Oil 30 | SAE 30 |
| TEXACO | Compressor Oil EP VDL 100 | VDL-100 |
| | Regal R&O 100 | VCL-100 |
| | Ursatex 30 | SAE 30 |
| | DORO AR 30 | SAE 30 |
| TOTAL | Dacnis P 100 | VDL-100 |
| | Carprano TD 30 | SAE 30 |
| | Milcano TC 30 | SAE 30 |
| | Disola M 3015 | SAE 30 |

Table 8.1b: List of typical lubricant suppliers and grades (Flushing Oils)

| Brand | Name |
|---|---|
| Agip | Rustica C SAE 30 |
| ARAL | Konit Motor Oil SAE 30 |
| AVIA | MK 1540 S |
| | Avilub MK 3000 |
| BP | MEK 20 W-20 |
| Castrol | Running-in and Preservation Oil |
| DEA | Dearnot EKM 642 SAE 30 |
| ELF | Stockage 30 |
| Esso | MZK Motor Oil HD 30 |
| | Antirust MZ 110 |
| FINA | Rusan NF Motor Oil SAE 30 |
| Mobil | Mobilarma 524 |
| Shell | Ensis Motor Oil 30 |
| TEXACO | Engine Oil EKM 146 SAE 30 |

# Chapter 9

# MAINTENANCE AND TROUBLESHOOTING

## 9.0 General

It goes without saying that preventative planned maintenance is the correct approach for the upkeep of any reciprocating air compressor. In the near future condition monitoring similar to that presently seen on screw units will play an increasing role in the maintenance of reciprocating air compressors used for general service and control air purposes.

To carry out the maintenance of the compressor correctly it is essential that the following points are referred to.

## 9.1 Operating and Maintenance Handbook

It is necessary that the operating and maintenance handbook for each type of compressor is studied carefully before attempting maintenance work on a unit which has not previously been overhauled by the operator even if a procedure is considered to be straightforward. As mentioned earlier, a number of newly designed air compressors of the 90deg 'V' and 'W' configuration have been introduced by the manufacturer's of air and water cooled compressors. These units have improved reliability and increased component life and therefore more attention must be paid to the procedures to obtain the maximum benefits of the new designs.

To give operators a comparison of the improvements that have been made in the reliability of the latest designs of air compressors compared to older designs a set of tables are presented at the end of the chapter (see Tables 9.1-9.5).

## 9.2 Running Conditions

When new air compressors have been 'run-in' it is important that the engineer in charge of the unit maintains records all the interstage and final stage air temperatures and pressures, crankcase oil pressure and vibration levels etc, under as many varying operating conditions as possible. For example, from North Atlantic to the tropics will give widely differing operational parameters.

To assist in the logging of compressor running hours it is important that an hours run meter is fitted in the starter/control panel for each compressor.

These conditions should be logged and full information documented so that other engineers who may take over the maintenance duties are fully aware of the operating and maintenance parameters of each compressor. Good maintenance records are an essential part of any compressor reliability.

## 9.3 Cleanliness

Machines should be kept free from both dust and oil whilst in normal service or on standby since a clean machine will allow observations to be made at an early stage and help identify if any joints are leaking etc.

Care must also be taken when the air compressor is opened up to make an inspection and carry out maintenance procedures to ensure that any dirt or debris does not enter the machine which may result in damage at a later date. Open ports ie, the air inlet to the compressor should be covered up to ensure that dirt and debris does not enter.

## 9.4 Unit Lifting

It is very important that the manufacturer's lifting instructions as detailed in the instruction manual are followed to lift either the compressor, motor or the complete unit.

## 9.5 Safety

Before commencing any maintenance work on the unit it is very important that all of the power supplies to the machine have been switched off and that the machine is completely isolated from any electrical contacts. Notices should be hung from a prominent place ie, on the control panel door stating that the compressor unit is under repair or out of order and that the unit had been isolated. In addition both the delivered air outlet valve and the water inlet (if this is the cooling medium) and outlet stop valves should be shut off.

If the diesel engine used with the emergency air compressor unit is also under maintenance a notice should also be hung from the starting handle. If the engine has a battery start the terminals should be disconnected.

When working in machinery spaces it is important that the ships' engineers should wear appropriate overalls, safety boots and hard hat. If paint or grinding work is taking place then safety goggles and mask should also be worn.

After a period of maintenance all equipment that has been used should be cleaned and stored in its appropriate place.

## 9.6 Lubrication Oil

The use of the correct grade of lubricating oil in the compressor is an important item that has to be selected to obtain maximum component life. There are a wide range of oils available today ranging from mineral to synthetic oils. The operator should always use oil according to the manufacturer's recommendations as laid down in the operations manual. Further details with reference to oils can be found in Chapter 8.

## 9.7 Correct Spare Parts

It is no secret that the shipowner can obtain his spare parts either from the original manufacturer or from the grey market and it is up to the shipowner which policy and risk that he will adopt in selecting the spare parts used. When an order is placed it is normally stated that the company must operate to the International Quality Standard ISO 9000 which in turn specifies that the parts supplied must be 'original part' only.

With the use of grey market parts with unidentified material specification no guarantee for compatibility with other parts (or for being to the latest design and modification), can be given. In addition the type approval and certificates of the selected classification society are required for starting air compressors which incorporate original parts. Modification of the compressor such as the fitting of non-authorized spare parts may invalidate the manufacturer's approval and guarantee and could lead to third party consequential damage.

## 9.8 Measuring Instruments

Any measuring instruments should be checked at regular intervals and recalibrated in accordance with, eg, British Standards and The Safety at Work Act etc.

## 9.9 Major Component Failure

In the event of a major component failure the operational data should be checked and if the problem cannot be identified the user should contact the manufacturer for his advice giving maximum details including operating conditions. To assist in a speedy conclusion to the problem the machine type and serial number identified on the nameplate on the side of the compressors should be given.

Routine checks should be carried out on maintenance sheets to ensure that any maintenance requirements have not been neglected.

## 9.10 Trouble Shooting

No matter how good the design, because of varying operating conditions problems may arise during the running of a reciprocating air compressor and it is essential that they can be identified and corrected quickly. Many of the problems can be found at an early stage by observing either manually or automatically crankcase oil pressure, interstage air pressures and operating temperatures. Tables 9.6 and 9.7 at the end of this Chapter show, respectively, lists of possible faults that may occur with a reciprocating air compressor and a screw compressor.

Table 9.1: Maintenance plan for the older type Vertical, Water Coo ed Reciprocating Compressor

| Maintenance | Runnning-in Work-hours | | | Period | During normal service | | | | | | | |
|---|---|---|---|---|---|---|---|---|---|---|---|---|
| | 10 | 20 | 50 | Daily | 500 | 1000 | 1500 | 2000 | 2500 | 3000 | 3500 | 4000 |
| Check oil and air pressure settings | x | x | x | | | | | | | | | |
| Check oil level | | | | x | | | | | | | | |
| Replace oil | | | | | | x | | x | | x | | x |
| Check air filter | | | | | x | | x | | x | | x | |
| 1st stage valve | | | | | x | x1 | x | x1 | x | x1 | x | x1 |
| 2nd stage valve | | | | | x | | x | x1 | x | x1 | x | x1 |
| Bearings | | | | | | x | | x | | | | x |
| Piston rings | | | | | | | | x | | | | x |
| Corrosion rods | | | | | | x | | x | | x | | x |
| Fusible plug | | | | | | x | | x | | x | | x |
| Coolers | | | | | | | | x | | | | x |
| Oil filter | | | | | | x | | x | | x | | x |
| Water pump (if fitted) | | | | | | x | | x | | x | | x |
| Drive belt tension | | | | | x | x | x | x | x | x | x | x |
| Unloaders | | | | | x | x | x | x | x | x | x | x |
| Safety devices | | | | | x | x | x | x | x | x | x | x |

Note: x = check or replace x1 = replace valves and springs

Table 9.2: Maintenance plan for a modern 90deg Vee Water Cooled Reciprocating Compressor

| Maintenance Work | Standard | Recommended | Designation | Part. No. | Qty. 500 h | Qty. 1000 h | Qty. 1500 h | Qty.2000 h | Qty. 2500 h | Qty. 3000 h | Qty. 3500 h | Qty. 4000 h | Qty. 4500 h | Qty. 5000 h | Qty. 5500 h | Qty. 6000 h | Qty. 6500 h | Qty. 7000 h | Qty. 7500 h | Qty. 8000 h | Qty. total |
|---|---|---|---|---|---|---|---|---|---|---|---|---|---|---|---|---|---|---|---|---|---|
| | | | Routine (8000 h) | | A | B | A | C | A | B | A | D | A | B | A | C | A | B | A | D | |
| R1 | × | | If necessary order spare parts for concentric/ lamellar valve according to spare parts list (or see R14/R4) | | | | | | | | | | | | | | | | | | |
| R1/R4/R9/R10/ R12 | × | | packing for air pipes of Cylinder head | | 1 | 1 | 1 | 1 | 1 | 1 | 1 | 1 | 1 | 1 | 1 | 1 | 1 | 1 | 1 | 1 | 16 |
| R1/R3/R7/R9/ R10/R11 | × | | o-ring for valve cover I. stage | | 2 | 2 | 2 | 2 | 2 | 2 | 2 | 2 | 2 | 2 | 2 | 2 | 2 | 2 | 2 | 2 | 32 |
| R1/R3/R7/R9/ R10/R11 | × | | o-ring for valve cover I. stage | | 2 | 2 | 2 | 2 | 2 | 2 | 2 | 2 | 2 | 2 | 2 | 2 | 2 | 2 | 2 | 2 | 32 |
| R1/R4/R9/R10/ R11 | × | | o-ring for valve cover II. stage | | 1 | 1 | 1 | 1 | 1 | 1 | 1 | 1 | 1 | 1 | 1 | 1 | 1 | 1 | 1 | 1 | 16 |
| R1/R4/R9/R10/ R11 | × | | o-ring for valve cover II. stage | | 1 | 1 | 1 | 1 | 1 | 1 | 1 | 1 | 1 | 1 | 1 | 1 | 1 | 1 | 1 | 1 | 16 |
| R2 | × | | If necessary order spare parts for dirt trap accord- -ing to spare parts list | | | | | | | | | | | | | | | | | | |
| R4 | × | | lamellar valve for II. stage | | | 1 | | 1 | | 1 | | 1 | | 1 | | 1 | | 1 | | 1 | 8 |
| R5 | × | | oil | | | 1 | | 1 | | 1 | | 1 | | 1 | | 1 | | 1 | | 1 | 8 |
| R5 | × | | cock gasket | | | 1 | | 1 | | 1 | | 1 | | 1 | | 1 | | 1 | | 1 | 8 |
| R6 | × | | If necessary order spare parts for air filter insert according to spare parts list (or see R15) | | | | | | | | | | | | | | | | | | |
| R7 | × | | concentric valve for I. stage | | | | | 2 | | | | 2 | | | | 2 | | | | 2 | 8 |

| | | | | | | | | | | | | | | | | | | | | | |
|---|---|---|---|---|---|---|---|---|---|---|---|---|---|---|---|---|---|---|---|---|---|
| R8 | × | | spare parts kit<br>Only valid for units with cooling water pump! | | | | | 1 | | | | 1 | | | | 1 | | | | 1 | 4 |
| R9/R10/R11 | × | | packing<br>for inspections hole | | | | | | | | | | | | | | | | | 2 | 2 |
| R9 | × | | R-ring for I. stage | | | | | | | | | | | | | | | | | 4 | 4 |
| R9 | × | | G-ring for I. stage | | | | | | | | | | | | | | | | | 2 | 2 |
| R9 | × | | R-ring for II. stage | | | | | | | | | | | | | | | | | 2 | 2 |
| R9 | × | | N-ring for II. stage | | | | | | | | | | | | | | | | | 1 | 1 |
| R9 | × | | S-ring for II. stage | | | | | | | | | | | | | | | | | 1 | 1 |
| R10 | × | | piston pin<br>for I. stage | | | | | | | | | | | | | | | | | 2 | 2 |
| R10 | × | | circlip<br>for I. stage | | | | | | | | | | | | | | | | | 2 | 2 |
| R10 | × | | piston pin<br>for II. stage | | | | | | | | | | | | | | | | | 1 | 1 |
| R10 | × | | circlip<br>for II. stage | | | | | | | | | | | | | | | | | 1 | 1 |
| R10 | × | | piston pin bearing<br>for I. stage | | | | | | | | | | | | | | | | | 2 | 2 |
| R10 | × | | piston pin bearing<br>for II. stage | | | | | | | | | | | | | | | | | 1 | 1 |
| R11 | × | | If necessary order spare parts for conecting rod bearings according to spare parts list<br>(or see R16) | | | | | | | | | | | | | | | | | | |
| R12 | × | | If necessary order spare parts for pistons and cylinder liners according to spare parts list | | | | | | | | | | | | | | | | | | |
| R13 | × | | If necessary order spare parts for flexible gear rim according to spare parts list (or see R17) | | | | | | | | | | | | | | | | | | |

Table 9.2 (Continued): Maintenance plan for a modern 90deg Vee Water Cooled Reciprocating Compressor

Table 9.2 (Continued): Maintenance plan for a modern 90deg Vee Water Cooled Reciprocating Compressor

| Routine (8000 h) | | | | | A | B | A | C | A | B | A | C | A | B | A | C | A | B | A | D | |
|---|---|---|---|---|---|---|---|---|---|---|---|---|---|---|---|---|---|---|---|---|---|
| Maintenance Work | Standard | Recommended | Designation | Part. No. | Qty. 500 h | Qty. 1000 h | Qty. 1500 h | Qty.2000 h | Qty. 2500 h | Qty. 3000 h | Qty. 3500 h | Qty. 4000 h | Qty. 4500 h | Qty. 5000 h | Qty. 5500 h | Qty. 6000 h | Qty. 6500 h | Qty. 7000 h | Qty. 7500 h | Qty. 8000 h | Qty. total |
| R14 | | × | suction valve plate for I. stage | | | 2 | | | | 2 | | | | 2 | | | | 2 | | | 8 |
| R14 | | × | delivery valve plate for I. stage | | | 2 | | | | 2 | | | | 2 | | | | 2 | | | 8 |
| R14 | | × | delivery valve plate for I. stage | | | 2 | | | | 2 | | | | 2 | | | | 2 | | | 8 |
| R14 | | × | delivery valve spring for I. stage | | | 4 | | | | 4 | | | | 4 | | | | 4 | | | 16 |
| R14 | | × | delivery valve spring for I. stage | | | 2 | | | | 2 | | | | 2 | | | | 2 | | | 8 |
| R14 | | × | suction valve spring for I. stage | | | 2 | | | | 2 | | | | 2 | | | | 2 | | | 8 |
| R15 | | × | filter insert for air filter | | | | | | | | | | | | | | | | | 2 | 2 |
| R16 | | × | connecting rod bearing | | | | | | | | | | | | | | | | | 3 | 3 |
| R17 | | × | flexible gear rim | | | | | | | | | | | | | | | | | 1 | 1 |
| R18 | | × | cylinder roller bearing (main bearing) | | | | | | | | | | | | | | | | | 1 | 1 |
| R18 | | × | shaft seal (main bearing) | | | | | | | | | | | | | | | | | 1 | 1 |
| R18 | | × | housing cover packing (main bearing) | | | | | | | | | | | | | | | | | 1 | 1 |

| Designation of Maintenance Works | | | |
|---|---|---|---|
| R1 (Routine A) | × | | check and clean concentric valve of I. stage and lamellar valve of II. stage |
| R2 (Routine A) | × | | only valid for compressors with cooling water control valves: cleaning of the dirt trap |
| R3 (Routine B) | × | | check concentric valve of I. stage, if necessary change plates and springs |
| R4 (Routine B) | × | | change lamellar valve of II. stage |
| R5 (Routine B) | × | | oil change |

| | | | |
|---|---|---|---|
| R6 (Routine B) | × | | check air filter insert, if necessary change them |
| R7 (Routine C) | × | | change concentric valve of I. stage |
| R8 (Routine C) | × | | only valid for compressors with cooling water pump: check cooling water pump, if necessary change parts |
| R9 (Routine D) | × | | change piston rings of all stages |
| R10 (Routine D) | × | | change piston pins and piston pin bearings of all stages |
| R11 (Routine D) | × | | check connecting rod bearings of all stages |
| R12 (Routine D) | × | | check pistons and cylinder liners, if necessary change them |
| R13 (Routine D) | × | | check elastic coupling, if necessary change flexible gear rim |
| R14 (Routine B) | | × | change plates and springs of concentric valve I. stage |
| R15 (Routine D) | | × | change air filter insert |
| R16 (Routine D) | | × | change connecting rod bearings of all stages |
| R17 (Routine D) | | × | change flexible gear rim |
| R18 (Routine D) | | × | change main bearings (cylinder roller bearing) |

Table 9.2 (Continued): Maintenance plan for a modern 90deg Vee Water Cooled Reciprocating Compressor

Table 9.3: Maintenance plan for a 90deg Vee, two-stage Air Cooled Reciprocating Compressor

| | | | Routine | | A | B | A | C | A | B | A | D | A | B | A | C | A | B | A | D | A | B | A | C | A | B | A | D | |
|---|---|---|---|---|---|---|---|---|---|---|---|---|---|---|---|---|---|---|---|---|---|---|---|---|---|---|---|---|---|
| Maintenance Work change of: | Standard | Recommended | Designation | Part. No. | Qty. 1000 h | Qty. 2000 h | Qty. 3000 h | Qty. 4000 h | Qty. 5000 h | Qty. 6000 h | Qty. 7000 h | Qty. 8000 h | Qty. 9000 h | Qty. 10000 h | Qty. 11000 h | Qty. 12000 h | Qty. 13000 h | Qty. 14000 h | Qty. 15000 h | Qty. 16000 h | Qty. 17000 h | Qty. 18000 h | Qty. 19000 h | Qty. 20000 h | Qty. 21000 h | Qty. 22000 h | Qty. 23000 h | Qty. 24000 h | Qty. total |
| connecting rod bearings disk | × | | circlip | | | | | | | | | 1 | | | | | | | | 1 | | | | | | | | 1 | 3 |
| piston pin | × | | circlip 1st stage | | | | | 2 | | | | 2 | | | | 2 | | | | 2 | | | | 2 | | | | 2 | 12 |
| piston pin | × | | circlip 2nd stage | | | | | 2 | | | | 2 | | | | 2 | | | | 2 | | | | 2 | | | | 2 | 12 |
| connecting rod bearings | × | | connecting bearing 1st/2nd stage | | | | | | | | | 2 | | | | | | | | 2 | | | | | | | | 2 | 6 |
| pistons, cylinders | × | | cylinder 1st stage | | | | | | | | | 1 | | | | | | | | 1 | | | | | | | | 1 | 3 |
| pistons, cylinders | × | | cylinder 2nd stage | | | | | | | | | 1 | | | | | | | | 1 | | | | | | | | 1 | 3 |
| routine B, C, D | × | | cylinder foot packing 1st stage | | | 1 | | 1 | | 1 | | 1 | | 1 | | 1 | | 1 | | 1 | | 1 | | 1 | | 1 | | 1 | 12 |
| routine B, C, D | × | | cylinder foot packing 2nd stage | | | 1 | | 1 | | 1 | | 1 | | 1 | | 1 | | 1 | | 1 | | 1 | | 1 | | 1 | | 1 | 12 |
| routine A, B, C, D | × | | cylinder head packing 1st stage | | 1 | 1 | 1 | 1 | 1 | 1 | 1 | 1 | 1 | 1 | 1 | 1 | 1 | 1 | 1 | 1 | 1 | 1 | 1 | 1 | 1 | 1 | 1 | 1 | 12 |
| routine A, B, C, D | × | | cylinder head packing 2nd stage | | 1 | 1 | 1 | 1 | 1 | 1 | 1 | 1 | 1 | 1 | 1 | 1 | 1 | 1 | 1 | 1 | 1 | 1 | 1 | 1 | 1 | 1 | 1 | 1 | 12 |
| main bearings | × | | cylinder roller bearing | | | | | | | | | 1 | | | | | | | | 1 | | | | | | | | 1 | 3 |
| main bearings | × | | cylinder roller bearing | | | | | | | | | 1 | | | | | | | | 1 | | | | | | | | 1 | 3 |
| flexible gear rim | × | | flexible gear rim f.couplg. | | | | | 1 | | | | 1 | | | | 1 | | | | 1 | | | | 1 | | | | 1 | 6 |
| piston rings | × | | G-ring 1st stage | | | | | | | | | 1 | | | | | | | | 1 | | | | | | | | 1 | 3 |
| oil | × | | gasket | | 1 | 1 | 1 | 1 | 1 | 1 | 1 | 1 | 1 | 1 | 1 | 1 | 1 | 1 | 1 | 1 | 1 | 1 | 1 | 1 | 1 | 1 | 1 | 1 | 24 |
| lamellar valves | × | | lamellar 1st stage | | | 1 | | 1 | | 1 | | 1 | | 1 | | 1 | | 1 | | 1 | | 1 | | 1 | | 1 | | 1 | 12 |
| lamellar valves | × | | lamellar valve 2nd stage | | | 1 | | 1 | | 1 | | 1 | | 1 | | 1 | | 1 | | 1 | | 1 | | 1 | | 1 | | 1 | 12 |
| lamellar valves | × | | low tolerance gasket 1st stage | | 1 | 1 | 1 | 1 | 1 | 1 | 1 | 1 | 1 | 1 | 1 | 1 | 1 | 1 | 1 | 1 | 1 | 1 | 1 | 1 | 1 | 1 | 1 | 1 | 24 |
| piston rings | × | | M-ring 1st stage | | | | | | | | | 1 | | | | | | | | 1 | | | | | | | | 1 | 3 |
| piston rings | × | | N-ring 1st stage | | | | | | | | | 1 | | | | | | | | 1 | | | | | | | | 1 | 3 |
| piston rings | × | | N-ring 2nd stage | | | | | | | | | 1 | | | | | | | | 1 | | | | | | | | 1 | 3 |
| lamellar valves | × | | O-ring 2nd stage | | 1 | 1 | 1 | 1 | 1 | 1 | 1 | 1 | 1 | 1 | 1 | 1 | 1 | 1 | 1 | 1 | 1 | 1 | 1 | 1 | 1 | 1 | 1 | 1 | 24 |

| | | | | | | | | | | | | | | | | | | | | | | | | | | | | | |
|---|---|---|---|---|---|---|---|---|---|---|---|---|---|---|---|---|---|---|---|---|---|---|---|---|---|---|---|---|---|
| routine D | × | | packing for bearing bracket | | | | | | | | | 1 | | | | | | | | 1 | | | | | | | | 1 | 3 |
| pistons, cylinders | × | | piston 1st stage | | | | | | | | | 1 | | | | | | | | 1 | | | | | | | | 1 | 3 |
| pistons, cylinders | × | | piston 2nd stage | | | | | | | | | 1 | | | | | | | | 1 | | | | | | | | 1 | 3 |
| piston pin | × | | piston pin 1st stage | | | | | 1 | | | | 1 | | | | 1 | | | | 1 | | | | 1 | | | | 1 | 6 |
| piston pin | × | | piston pin 2nd stage | | | | | 1 | | | | 1 | | | | 1 | | | | 1 | | | | 1 | | | | 1 | 6 |
| piston pin | × | | piston pin bearing 1st stage | | | | | 1 | | | | 1 | | | | 1 | | | | 1 | | | | 1 | | | | 1 | 6 |
| piston pin | × | | piston pin bearing 2nd stage | | | | | 1 | | | | 1 | | | | 1 | | | | 1 | | | | 1 | | | | 1 | 6 |
| piston rings | × | | R-ring 2nd stage | | | | | | | | | 2 | | | | | | | | 2 | | | | | | | | 2 | 6 |
| piston rings | × | | S-ring 2nd stage | | | | | | | | | 1 | | | | | | | | 1 | | | | | | | | 1 | 3 |
| shaft steals | × | | shaft steal | | | | | | | | | 1 | | | | | | | | 1 | | | | | | | | 1 | 3 |
| shaft steals | × | | shaft steal | | | | | | | | | 1 | | | | | | | | 1 | | | | | | | | 1 | 3 |
| connecting rod bearings disk | × | | washer | | | | | | | | | 1 | | | | | | | | 1 | | | | | | | | 1 | 3 |
| air filter insert | | × | filter insert for air filter | | 1 | 1 | 1 | 1 | 1 | 1 | 1 | 1 | 1 | 1 | 1 | 1 | 1 | 1 | 1 | 1 | 1 | 1 | 1 | 1 | 1 | 1 | 1 | 1 | 24 |
| piston rings | | × | G-ring 1st stage | | | | | 1 | | | | | | | | 1 | | | | | | | | 1 | | | | | 3 |
| safety valves | | × | gasket | | | | | | | | | 1 | | | | | | | | 1 | | | | | | | | 1 | 3 |
| lamellar valves | | × | lamellar valve 1st stage | | 1 | | 1 | | 1 | | 1 | | 1 | | 1 | | 1 | | 1 | | 1 | | 1 | | 1 | | 1 | | 12 |
| lamellar valves | | × | lamellar valve 2nd stage | | 1 | | 1 | | 1 | | 1 | | 1 | | 1 | | 1 | | 1 | | 1 | | 1 | | 1 | | 1 | | 12 |
| piston rings | | × | M-ring 1st stage | | | | | 1 | | | | | | | | 1 | | | | | | | | 1 | | | | | 3 |
| piston rings | | × | N-ring 1st stage | | | | | 1 | | | | | | | | 1 | | | | | | | | 1 | | | | | 3 |
| piston rings | | × | N-ring 2nd stage | | | | | 1 | | | | | | | | 1 | | | | | | | | 1 | | | | | 3 |
| piston rings | | × | R-ring 2nd stage | | | | | 2 | | | | | | | | 2 | | | | | | | | 2 | | | | | 6 |
| piston rings | | × | S-ring 2nd stage | | | | | 1 | | | | | | | | 1 | | | | | | | | 1 | | | | | 3 |
| safety valves | | × | safety valves 1st stage | | | | | | | | | 1 | | | | | | | | 1 | | | | | | | | 1 | 3 |
| safety valves | | × | safety valves 2nd stage | | | | | | | | | 1 | | | | | | | | 1 | | | | | | | | 1 | 3 |

Table 9.3 (Continued): Maintenance plan for a 90deg Vee, two-stage Air Cooled Reciprocating Compressor

Table 9.4: Maintenance plan for a W ' configuration, three stage Air Cooled Reciprocating Compressor

| | | | *Routine (24000 h)* | | A | B | A | C | A | B | A | D | A | B | A | C | A | B | A | D | A | B | A | C | A | B | A | D | |
|---|---|---|---|---|---|---|---|---|---|---|---|---|---|---|---|---|---|---|---|---|---|---|---|---|---|---|---|---|---|
| *Maintenance Work* | Standard | Recommended | Designation | Part. No. | *Qty. 1000 h* | *Qty. 2000 h* | *Qty. 3000 h* | *Qty. 4000 h* | *Qty. 5000 h* | *Qty. 6000 h* | *Qty. 7000 h* | *Qty. 8000 h* | *Qty. 9000 h* | *Qty. 10000 h* | *Qty. 11000 h* | *Qty. 12000 h* | *Qty. 13000 h* | *Qty. 14000 h* | *Qty. 15000 h* | *Qty. 16000 h* | *Qty. 17000 h* | *Qty. 18000 h* | *Qty. 19000 h* | *Qty. 20000 h* | *Qty. 21000 h* | *Qty. 22000 h* | *Qty. 23000 h* | *Qty. 24000 h* | *Qty. total* |
| R1 | × | | oil | | 1 | 1 | 1 | 1 | 1 | 1 | 1 | 1 | 1 | 1 | 1 | 1 | 1 | 1 | 1 | 1 | 1 | 1 | 1 | 1 | 1 | 1 | 1 | 1 | 24 |
| R1 | × | | gasket | | 1 | 1 | 1 | 1 | 1 | 1 | 1 | 1 | 1 | 1 | 1 | 1 | 1 | 1 | 1 | 1 | 1 | 1 | 1 | 1 | 1 | 1 | 1 | 1 | 24 |
| R2 | × | | If necessary order spare parts for air filter insert according to spare parts list (or see R13) | | | | | | | | | | | | | | | | | | | | | | | | | | |
| R3 | × | | If necessary order spare parts for concentric valve according to spare parts list (or see R12) | | | | | | | | | | | | | | | | | | | | | | | | | | |
| R3/R5/R6/R7/R8/ R10/R14/R16 | × | | cylinder head packing for I. stage | | | 2 | | 2 | | 2 | | 2 | | 2 | | 2 | | 2 | | 2 | | 2 | | 2 | | 2 | | 2 | 24 |
| R3/R5/R6/R7/R8 R10/R14/R16 | × | | low tolerance gasket (cylinder head) for I. stage | | | 1 | | 1 | | 1 | | 1 | | 1 | | 1 | | 1 | | 1 | | 1 | | 1 | | 1 | | 1 | 12 |
| R4 | × | | If necessary order spare parts for lamellar valves according to spare parts list (or see R9) | | | | | | | | | | | | | | | | | | | | | | | | | | |
| R4/R5/R6/R7/R9 R10/R14/R16 | × | | cylinder head packing for II. stage | | | 1 | | 1 | | 1 | | 1 | | 1 | | 1 | | 1 | | 1 | | 1 | | 1 | | 1 | | 1 | 12 |
| R4/R5/R6/R7/R9/ R10/R14/R16 | × | | cylinder head packing for III. stage | | | 1 | | 1 | | 1 | | 1 | | 1 | | 1 | | 1 | | 1 | | 1 | | 1 | | 1 | | 1 | 12 |
| R4/R5/R6/R7/R9/ R10/R14/R16 | × | | O-ring (cylinder head) for III. stage | | | 1 | | 1 | | 1 | | 1 | | 1 | | 1 | | 1 | | 1 | | 1 | | 1 | | 1 | | 1 | 12 |
| R5 | × | | If necessary order spare parts for piston rings according to spare parts list (or see R14) | | | | | | | | | | | | | | | | | | | | | | | | | | |
| R5/R6/R7/R10/ R14/R16 | × | | cylinder foot packing for I. stage | | | | | | | 1 | | | | | | 1 | | | | | | 1 | | | | | | 1 | 4 |

| | | | | | | | | | | | | | | | | | | | | | | | | | | | | | |
|---|---|---|---|---|---|---|---|---|---|---|---|---|---|---|---|---|---|---|---|---|---|---|---|---|---|---|---|---|---|
| R5/R6/R7/R10 R14/R16 | × | | cylinder foot packing for II. and III. stage | | | | | | | 2 | | | | | | 2 | | | | | | 2 | | | | | | 2 | 8 |
| R6 | × | | piston pin for I. stage | | | | | | | 1 | | | | | | 1 | | | | | | 1 | | | | | | 1 | 4 |
| R6 | × | | piston pin for II. stage and III. stage | | | | | | | 2 | | | | | | 2 | | | | | | 2 | | | | | | 2 | 8 |
| R6 | × | | circlip for all stages | | | | | | | 6 | | | | | | 6 | | | | | | 6 | | | | | | 6 | 24 |
| R6/R7/R16 | × | | packing for cover | | | | | | | 2 | | | | | | 2 | | | | | | 2 | | | | | | 2 | 8 |
| R6 | × | | piston pin bearing for all stages | | | | | | | 3 | | | | | | 3 | | | | | | 3 | | | | | | 3 | 12 |
| R7/R16 | × | | packing for bearing bracket | | | | | | | | | | | | | 1 | | | | | | | | | | | | 1 | 2 |
| R7/R16 | × | | shaft seal | | | | | | | | | | | | | 1 | | | | | | | | | | | | 1 | 2 |
| R7/R16 | × | | shaft seal | | | | | | | | | | | | | 1 | | | | | | | | | | | | 1 | 2 |
| R7 | × | | connecting rod bearing for all stages | | | | | | | | | | | | | 3 | | | | | | | | | | | | 3 | 6 |
| R8 | × | | concentric valve for I stage | | | | | | | | | | | | | 1 | | | | | | | | | | | | 1 | 2 |
| R9 | × | | lamellar valve for II. stage | | | | | | | | | | | | | 1 | | | | | | | | | | | | 1 | 2 |
| R9 | × | | lamellar valve for III. stage | | | | | | | | | | | | | 1 | | | | | | | | | | | | 1 | 2 |
| R10 | × | | If necessary order spare parts for piston or cylinder according to spare parts list | | | | | | | | | | | | | | | | | | | | | | | | | | |
| R11 | × | | If necessary order spare parts for flexible gear rim according to spare parts list (or see R15) | | | | | | | | | | | | | | | | | | | | | | | | | | |
| R12 | | × | suction valve plate for I. stage | | | 1 | | 1 | | 1 | | 1 | | 1 | | | | 1 | | 1 | | 1 | | 1 | | 1 | | | 10 |

Table 9.4 (Continued): Maintenance plan for a 'W' configuration, three stage Air Cooled Reciprocating Compressor

Table 9.4 (Continued): Maintenance plan for a 'W' configuration, three stage Air Cooled Reciprocating Compressor

| | | | Routine (24000 H) | | A | B | A | B | A | C | A | B | A | B | A | D | A | B | A | B | A | C | A | B | A | B | A | D | |
|---|---|---|---|---|---|---|---|---|---|---|---|---|---|---|---|---|---|---|---|---|---|---|---|---|---|---|---|---|---|
| Maintenance Work | Standard | Recommended | Designation | Part. No. | Qty. 1000 h | Qty. 2000 h | Qty. 3000 h | Qty. 4000 h | Qty. 5000 h | Qty. 6000 h | Qty. 7000 h | Qty. 8000 h | Qty. 9000 h | Qty. 10000 h | Qty. 11000 h | Qty. 12000 h | Qty. 13000 h | Qty. 14000 h | Qty. 15000 h | Qty. 16000 h | Qty. 17000 h | Qty. 18000 h | Qty. 19000 h | Qty. 20000 h | Qty. 21000 h | Qty. 22000 h | Qty. 23000 h | Qty. 24000 h | Qty. total |
| R12 | | × | suction valve spring for I. stage | | | 1 | | 1 | | 1 | | 1 | | 1 | | | | 1 | | 1 | | 1 | | 1 | | 1 | | | 10 |
| R12 | | × | delivery valve plate for I. stage | | | 2 | | 2 | | 2 | | 2 | | 2 | | | | 2 | | 2 | | 2 | | 2 | | 2 | | | 20 |
| R12 | | × | delivery valve spring for I. stage | | | 1 | | 1 | | 1 | | 1 | | 1 | | | | 1 | | 1 | | 1 | | 1 | | 1 | | | 10 |
| R12 | | × | suction valve plate for I. stage | | | 1 | | 1 | | 1 | | 1 | | 1 | | | | 1 | | 1 | | 1 | | 1 | | 1 | | | 10 |
| R13 | | × | filter insert for air filter | | | | | | | | | | | | | 1 | | | | | | | | | | | | 1 | 2 |
| R14 | | × | M-ring for I. stage | | | | | | | | | | | | | 1 | | | | | | | | | | | | 1 | 2 |
| R14 | | × | N-ring for I. stage | | | | | | | | | | | | | 1 | | | | | | | | | | | | 1 | 2 |
| R14 | | × | G-ring for I. stage | | | | | | | | | | | | | 1 | | | | | | | | | | | | 1 | 2 |
| R14 | | × | R-ring for II. stage | | | | | | | | | | | | | 1 | | | | | | | | | | | | 1 | 2 |
| R14 | | × | N-ring for II. stage | | | | | | | | | | | | | 1 | | | | | | | | | | | | 1 | 2 |
| R14 | | × | S-ring for II. stage | | | | | | | | | | | | | 1 | | | | | | | | | | | | 1 | 2 |
| R14 | | × | R-ring for III. stage | | | | | | | | | | | | | 2 | | | | | | | | | | | | 2 | 4 |
| R14 | | × | N-ring for III. stage | | | | | | | | | | | | | 1 | | | | | | | | | | | | 1 | 2 |
| R15 | | × | flexible gear rim | | | | | | | | | | | | | 1 | | | | | | | | | | | | 1 | 2 |
| R16 | | × | cylinder roller bearing (main bearing) | | | | | | | | | | | | | 2 | | | | | | | | | | | | 2 | 4 |

Designation of Maintenance Works

| | | | |
|---|---|---|---|
| R1 (Routine A) | × | | oil change |
| R2 (Routine A) | × | | check filter |
| R3 (Routine B) | × | | check concentric valve of I. stage, if necessary change plates and springs |
| R4 (Routine B) | × | | change lamellar valve of II. stage and III. stage, if necessary change valves |
| R5 (Routine C) | × | | change piston rings of all stages, if necessary change it |
| R6 (Routine C) | × | | change piston pins and pistons pin bearings of all stages |

| | | | |
|---|---|---|---|
| R7 (Routine D) | | × | change connecting rod bearings of all stages |
| R8 (Routine D) | × | | change concetric valve of I. stage |
| R9 (Routine D) | × | | change lamellar valves of II. stage and III. stage |
| R10 (Routine D) | × | | check piston and cylinder, if necessary change it |
| R11 (Routine D) | × | | check elastic coupling, if necessary change flexible gear rim |
| R12 (Routine B) | | × | change plates and springs of concentric valve |
| R13 (Routine D) | | × | change air filter |
| R14 (Routine D) | | × | change piston rings of all stages |
| R15 (Routine D) | | × | change flexible gear grim |
| R16 (Routine D) | | × | change main bearings (cylinder roller bearing) |

Table 9.4 (Continued): Maintenance plan for a 'W' configuration, three stage Air Cooled Reciprocating Compressor

Table 9.5: Service Activities for a Modern Air Cooled Screw Compressor

| Service activities | After the first 100 operating hours | After the first 500 operating hours | Every week | Every 500 operating hours | Every 2000 operating hours but at least once a year |
|---|---|---|---|---|---|
| Check cooling fluid level | × | | × | | |
| Check for leakage | × | | × | | |
| Check compressor temperature | × | | × | | |
| Check cooler for fouling | × | | × | | |
| Check for condensate | × | | × | | |
| Replace cooling fluid and cooling fluid filter | | × | | | × |
| Replace cooling liquid separator | | | | | × |
| Replace air filter | | | | | × |
| Check safety valve | | | | | × |
| Check drive | | | | × | |
| Lubrication of motor | | | | | ×* |
| General compressor servicing | | | | | × |

*The motor lubrication intervals vary between 3100 and 20 000 hours depending on the type of unit. A warning on the control indicates when the motor has to be lubricated.

## Table 9.6: Troubleshooting – possible faults with a Reciprocating Air Compressor

**Note!**

- In case of trouble first check the indicators of compressor control and compressor
- Try to remedy the fault by following the information given in the table below

| Fault | Likely cause | Remedy |
|---|---|---|
| Compressor does not start or switches OFF | No supply voltage/ no control voltage | Check fuses. Replace blown fuses |
| Oil pressure monitor has responded | The oil level is too low | Check the oil level, add oil as required |
| | The oil pump is faulty or leaking | Check the oil pump. Fix the leakage or replace the oil pump as required |
| Compressor was shut down by the overcurrent relay of the control system | The motor is overheated Excessive current drawn | Determine the cause of the fault and remedy it. The compressor can be started again after being allowed to cool off |
| | Piston seizure | Check cylinder and piston for score marks, replace if necessary |
| Safety valve of 1st stage blows off: | | |
| Pressure exceeds blowing-off pressure (8 bar) | 2nd stage valve is not working properly | Check 2nd stage valve, replace if necessary |
| | Sealing between inlet and outlet side of the 2nd stage is faulty | Replace gasket |
| Pressure below blowing-off pressure (8 bar) | Safety valve faulty | Replace safety valve |
| 2nd stage safety valve blows off: | | |
| Pressure above blowing-off pressure (ultimate pressure + 5%) | Valve in air line to compressed air receiver closed | Open the valve |
| | Pressure switch set too high | Reduce the set pressure |
| Pressure below blowing-off pressure (ultimate pressure + 5%) | Safety valve set too low or is faulty | Replace safety valve |
| Pressure in the 1st stage excessively high | 2nd stage valve leaking | Check valve for damage, and if necessary replace; replace gaskets |
| Pressure gauge of the 1st stage indicates lower pressure | 1st stage valve leaking | Check 1st stage valve, replace if necessary |
| | Air filter very dirty | Replace air filter cartridge |
| No pressure indicated in 1st and 2nd stage pressure gauges | No power at solenoid valve | Check power supply of solenoid valve |
| | Solenoid valve faulty | Check solenoid valve, replace if necessary |
| Air escaping from compressed air lines | Gaskets of connections leaking | Replace relevant gasket |
| | Compression ring joints leaky | Turn compressor OFF. Wait until all parts are relieved of pressure (check pressure gauges). Then tighten unions. |
| Temperature control has responded | Compressed air temperature at outlet too high due to cooling water deficiency | Increase cooling water flow-rate (only in case of the version with cooling water pump: open the gate valve behind the pump further), check the temperature of the cooling water at the outlet and the function of the cooling water stop valve; clean particle trap. |
| | Compressed air-temperature at the outlet too high due to poor heat transfer in the cooler as a result of boiler scale formation | Clean cooling water cavities with boiler cleaning compound |

Table 9.6 (continued): Troubleshooting – possible faults with a Reciprocating Air Compressor

| Fault | Likely cause | Remedy |
|---|---|---|
| Drain valve does not close | No supply voltage | Check fuses, replace blown fuses |
| | Solenoid faulty | Replace solenoid |
| | Foreign matter in solenoid valve | Clean solenoid valve |
| Cooling water stop valve does not open | No supply voltage | Check fuses, replace blown fuses |
| | Solenoid faulty | Replace solenoid |
| | Foreign matter in solenoid valve | Clean solenoid valve |
| Compressor makes loud noises | Connecting rod bearing faulty | Check connecting rod bearing, replace if necessary. Check oil supply. |
| | Gudgeon pin bearing faulty | Check gudgeon pin bearing, replace if necessary |
| | Crankshaft bearing faulty | Check crankshaft bearing, replace if necessary |
| Oil leaking from crankcase | Gasket or shaft seal faulty<br>Screws loosened | Tighten screws<br>**If there is heavy leaking**, check to see which gasket is faulty, then replace it<br>**Minor traces of oil** on the crankcase or oil drops below the compressor are harmless. Wipe off with a rag |
| Water is leaking at the relief groove of the cylinder flange surface | Liner o-ring above the relief groove is faulty | Replace the o-ring |
| Oil is leaking at the relief groove of the cylinder flange surface | Liner o-ring below the relief groove is faulty | Repalce the o-ring |
| Water in oil | Compressor is overcooled because of<br>-too high cooling water flow-rate<br>-improper room ventilation<br>-very short compressor running time | Reduce cooling water flow-rate. Change the room ventilation in such a way that it does not blow directly at the compressor; increase the compressor running time |
| | Insufficient drainage | Check drain lines and drain intervals |
| Burst disc in the cooling water circuit responds | Leakages in the cooler | Re-expansion of the cooler pipes in the cylinder housing. |
| | Pressure peaks in the cooling water circuit, which lead to the overstepping of the bursting pressure of the burst disc | Identify the causes for pressure peaks and remedy or reduce them (starting of additional stronger pumps, actuation of electromagnetic stop or alternating check valves or similar) |
| Untimely breaking or valve plates, valve springs, or valve discs | Insufficient drainage | Check drain lines and drain intervals<br>**Note:** Traces of impact on the sealing seat of the valve plate are normal |

Table 9.7: Troubleshooting – possible faults with a Screw Compressor

| Fault | Likely cause | Remedy |
| --- | --- | --- |
| Compressor does not start automatically or does not deliver air after previous switching off upon reaching the final pressure or from offload state | Net pressure set too high | Adjust net pressure anew |
| | Interruption in the control current circuit | Check electric circuit for interruption (only by trained electrician) |
| | Ambient temperature under +1°C; message **'Cooling fluid temperature too low'** | Install an additional heater or keep compressor room at right temperature |
| | Switching times have been activated in Air Control 3 | Check switching times and pressure times in Air Control |
| System does not start on pressing the start button (I) | Line pressure greater than start pressure | Note line pressure value |
| | Remote control activated<br>Missing voltage at the compressor | Symbol 'remote' is blinking<br>Check if voltage is applied |
| | Electrical error in the control system | Inspect (only by a trained electrician) |
| | Switching times have been activated in Air Control 3 | Check switching times in the Air Control 3 |
| Compressed air containing large amounts of cooling fluid (cooling fluid consumption too great) | Cooling fluid return flow piping is congested | Clean cooling fluid feedback piping |
| | Defective cooling fluid separator | Replace cooling fluid separator |
| System stopped before reaching the final pressure (red lamp is on) | Overtemperature or overpressure | Rectify error as required |
| | Interruption in the control current loop | Check current loop (only by a trained electrician) |
| Water in the piping net | Dryer switched off | Switch on dryer |
| | Condensation diverter is not functioning | Clean/exchange drain |
| | Bypass open | Close bypass |
| | Dew point too high | Demand customer service |
| Pressure decline | Pressure differential in the filter too great | Exchange filter |
| Compression temperature too high (red lamp is on) | Silencing hood not closed | Check and secure sound-insulation hood |
| | Intake or ambient temperature too high | Ventilate compressor room |
| | Cooling air inlet or outlet blocked | Make sufficient room |
| | Cooling fluid filter fouled<br>Insufficient cooling fluid | Renew cooling fluid filter<br>Add cooling fluid |
| | External cooling fluid heat exchanger fouled! **Attention:** Cooler screw should always be worked with a counter wrench; keep from applying torque to the cooler | Clean with compressed air. In the case of extensive fouling: disassemble cooler and clean with high pressure cleaner **Attention: Danger of short circuit!** Do not put electrical elements under power |

Table 9.7 (continued): Troubleshooting – possible faults with a Screw Compressor

| Fault | Likely cause | Remedy |
|---|---|---|
| Line pressure falls | Compressed air consumption greater than delivery quantity of the compressor | A compressor with larger delivery quantity is required |
| | Air filter is fouled | Replace air filter |
| | Relief valve blows air during compression | Check relief valve; if necessary, replace seals |
| | Suction regulator does not open | Check solenoid valve regulator spool and if necessary, replace |
| | Leakage in the pipework | Make pipework airtight |
| System pressure released by safety valve | Line pressure set too high | Adjust line pressure anew |
| | Safety valve defective | Check safety valve; if necessary, exchange |
| | Minimum pressure valve blocked | Exchange solenoid valve |
| | Cooling fluid separator cartridge fouled | Exchange cooling fluid separator cartridge |
| **'Malfunction over-pressure'** or **'line pressure too high'** (red lamp is on) | Cooling fluid separator fouled | Replace cooling fluid separator |
| | Higher external pressure in the compressed air net | Equalize outside pressure or remove from net |